Defects and Failures in Pressure Vessels and Piping

HELMUT THIELSCH

Vice President
Research, Development and Engineering
ITT Grinnell Corporation

ROBERT E. KRIEGER PUBLISHING COMPANY
HUNTINGTON, NEW YORK
1975

Original Edition 1965
Reprint 1977, with new material

Printed and Published by
ROBERT E. KRIEGER PUBLISHING CO., INC.
645 NEW YORK AVENUE
HUNTINGTON, NEW YORK 11743

Copyright © 1965 by
LITTON EDUCATIONAL PUBLISHING, INC.
Reprinted by Arrangement with
Van Nostrand Reinhold Co.

All rights reserved. No reproduction in any form of this book, in whole or in part (except for brief quotation in critical articles or reviews), may be made without written authorization from the publisher.

Printed in the United States of America

Library of Congress Cataloging in Publication Data

Thielsch, Helmut.
    Defects and failures in pressure vessels and piping.

    Reprint of the edition published by Reinhold Pub. Corp., New York.
    Includes bibliographies.
    1. Pipe.   2. Pressure vessels.   I. Title.
[TS280.T48   1975]   681'.76   75-15675
ISBN 0-88275-308-8

To

DR. ING. KURT THIELSCH

Preface

The materials used in pressure vessels, tanks, and piping, their fabrication, and the inspection of component parts or of the erected equipment normally are governed by a variety of standards and specifications. Quality control requirements cover all aspects of production from the time the base metal is produced to the time the erected vessel or piping system is placed in operation. During this time, the plate or pipe mill, foundry, fabricator and erector perform a multitude of operations, any one of which can result in a defective product if improperly performed.

Unfortunately, tests and inspection methods are often specified, the results of which are totally irrelevant insofar as confirming the suitability of the vessel or piping component for the intended service. A failure may then be the result.

In general, managements, purchasing departments, and many engineers have little appreciation of the factors and conditions leading to a failure. After an unexpected failure, the author is frequently rushed to a meeting where he is handed a piece of dirty metal and is asked "Why did it fail?" Disappointment reigns supreme when a spontaneous solution is not provided.

Failure analysis by a metallurgical specialist, in many respects, should be no different than the careful procedures followed by a medical specialist determining in a patient the causes of an unknown ache, pain or "flaw." All pertinent data and background information and environmental history should be gathered as basis for the subsequent careful examination. Extensive experience and a familiarity with the work done by other experts in related fields are also of utmost importance. This ensures, then, that the subsequent examination and testing follows in a logical order and involves the minimum number of tests to provide an accurate diagnosis.

Over the past three years, a series of articles have been published in *Heating, Piping and Air Conditioning* on "The Prevention of

Pipe Failures" involving primarily failures analyzed by the author. For the purpose of this book, this material has been greatly expanded and revised. Several major sections have been added. In addition, the subject matter has been broadened to include other references and failure experiences.

Information on defects and failures is scattered throughout a great many different publications. No attempt has been made to cover all the material published, or to list all references applicable to a particular defect, concept, or failure condition.

A great deal of important information and failure experience is also locked up in the engineering files of companies fearing adverse publicity of their operations or products. This, unfortunately, is shortsighted, since this unavailable experience could be of great benefit to design and engineering practices, material quality considerations, and fabrication and welding know-how.

Numerous ferrous and nonferrous materials are used in pressure vessels, tanks, and piping, singly or in combination. Defects are present in all of them, and failures have occurred in all major materials, and will occur again sooner or later. To illustrate all defects or show examples of failures applicable to every major material would be a perpetual project continuing to the end of time. Thus, only examples of typical defects and failures can be presented which apply to a few materials. The conditions applicable to other materials must then be interpreted. Actually, even in different materials, failure conditions are often very similar. Moreover, the same causes and failure mechanisms may apply just as much to a pipe section as to a pressure vessel or tank. In fact, cracking, crack propagation, corrosion, and erosion of the same materials often will look alike to the naked eye or under the microscope even though the section which failed was in a bridge, ship, building, airplane, missile or pressure vessel. Thus, much of what will be discussed and illustrated in this book will apply to all major engineering materials and structures.

The author hopes that this book, containing some previously unpublished information, will contribute in a small way to a recognition of the conditions leading to service failures. By utilizing this information in the establishment of realistic quality control requirements, our future industrial progress may be furthered.

As Vice-Admiral H. G. Rickover stated in his Keynote Address before the 44th Annual National Metal Congress on October 29, 1962, "I submit we must progress, and so we must pay the price of progress. We must accept the inexorably rising standards of technology, and we must relinquish comfortable routines and practices rendered obsolete because they no longer meet the new standards. This is our never-ending challenge."

HELMUT THIELSCH

Providence, Rhode Island
April, 1965

Acknowledgments

The author wishes to express his sincere gratitude to the numerous individuals with whom he has been involved in the examination and study of service failures. These include the organizations of clients as well as other groups with whom he has been associated. Many individuals in other companies have also contributed comments, helpful information, samples of failures, and constructive suggestions. Because of the subject matter involved, a number of these individuals have requested that their contributions remain anonymous.

The author is also particularly indebted to his associates A. Laurenson, E. M. Phillips, E. Johnson, R. Scott, S. Hawe, T. Martin, C. Ey and N. Kay, who have done much of the tedious and unglamorous detail work so necessary in defect analyses and failure examinations.

Finally, the author is deeply grateful to Robert W. Roose and Donald R. Bahnfleth, Editors of *Heating, Piping, and Air Conditioning* magazine for their continuous encouragement and support in the preparation of the initial series of 12 articles. Without their understanding and gently applied pressure on 12 magazine publication deadlines, the original articles serving as basis for this extensively revised and expanded manuscript would not have passed the idea stage.

Contents

Preface	v
CHAPTER 1 INTRODUCTION	1
Definitions	1
Laboratory Test Evaluation of Defects	2
Code Recognition of Defects	3
Inspection for Defects	3
Brittle and Ductile Failures	4
References	6
CHAPTER 2 CAUSES OF SERVICE FAILURES	7
Classifications	7
Failure Mechanism	8
Crack Initiation	9
Crack Growth	10
Crack Propagation	10
CHAPTER 3 DESIGN CONSIDERATIONS	12
Flexibility	12
Bends and Miters	17
Tanks	17
Design Notches	17
Sharp Corners	17
Changes in Wall Thickness	19
Outlets	23
Stress-concentration Factors	26
Nozzles at Elevated Temperatures	29
Attachments and Supports	29
Weld-Joint Design	35
Backing Rings	35
Elimination of Backing Rings	37
Proper Weld-joint Design	41

CHAPTER 4 MATERIAL SELECTION — 49

GENERAL CONSIDERATIONS — 49
LABORATORY TEST RESULTS — 50
CARBON STEEL MATERIALS — 51
GRAPHITE FORMATION, EMBRITTLEMENT AND CRACKING — 53
    Service Experience — 53
    Effects of Steel-making Practice — 57
    Metallographic Grading of Graphitization — 58
    Rehabilitation of Graphitized Weld Areas — 62
    Current Practice — 64
FERRITIC ALLOY STEELS IN HIGH-TEMPERATURE SERVICE — 64
    Vanadium Steels — 64
STAINLESS STEEL FOR CORROSION SERVICE — 66
    Failures in Columbium-bearing Grades — 67
    Sigma Phase in Stainless Steels — 70
    Simulated Service Tests at Elevated Temperatures — 72
SELECTION OF WELDING FILLER METAL — 72
DISSIMILAR JOINTS — 73
    Slight Dissimilarities in Composition — 73
    Potential Failures in Slightly Dissimilar Joints in Elevated Temperature Service — 74
    Major Dissimilarities in Composition — 74
    Factors Causing Failure — 75
    Elevated Temperature Service — 76
PREFABRICATED DISSIMILAR TRANSITION SECTIONS — 79
REFERENCES — 80

CHAPTER 5 IMPROPER HANDLING AND PROCESSING — 84

HANDLING OF PLATES, PIPING AND EQUIPMENT — 84
TRANSPORTATION OF VESSELS AND PIPE — 85
PROCESSING OF WELDING ELECTRODES — 86
STORAGE OF WELDING ELECTRODES — 87
REFERENCES — 87

CHAPTER 6 MATERIAL IDENTIFICATION — 88

INCORRECT IDENTIFICATION — 88
    Base Metals — 88
    Welding Filler Metals — 90
    Absence of Identification — 91
CARELESS ERECTION PRACTICE — 94
REFERENCES — 94

CHAPTER 7 DEFECTS IN WROUGHT AND FORGED PRODUCTS 96

NOTCH CONDITIONS 96
MECHANICAL NOTCHES 97
Laminations 98
Laps 100
Slivers 105
Scabs 105
Seams 107
Bark 107
Cracks 109
Code Acceptability Limits 111
Forging Defects 113
Bar Stock Defects 113
Tool Marks and Gouges 113
Notches in Welded Pipe Seams 116
METALLURGICAL NOTCHES 117
Hot Shortness 117
Surface Carburization and Decarburization 123
Hardness Variations 128
REFERENCES 128

CHAPTER 8 CASTING DEFECTS 130

MECHANICAL NOTCHES 130
Gas and Blow Holes 131
Inclusions 131
Shrinkage Cavities 132
Test Results 136
Service Experience 136
Hot Tears and Cracks 137
Unfused Chaplet 139
METALLURGICAL NOTCHES 140
Dissimilar Metal Chills 140
Hot Shortness 142
Hardness Variations 142
WELD REPAIRS 143
REFERENCES 146

CHAPTER 9 NOTCH BRITTLENESS 147

SERVICE FAILURES 148
NOTCH BRITTLENESS 150

DUCTILE VS BRITTLE BEHAVIOR	151
Characteristics	151
Low-temperature Service	155
Nature of Brittle Behavior	155
Evaluation of Brittle Behavior	156
Effects of Temperature	159
Transition Temperature	159
Effects of Notches and Notch Sensitivity	161
Fracture Analysis Diagram	164
Isothermal Crack Arrest Concept	167
TEST METHODS	168
Small Scale Test Specimens	168
Code-test requirements	172
Significance of Test Data	174
Ductility and Fracture Transition	174
FACTORS DETERMINING TRANSITION TEMPERATURES	176
Metallurgical Factors	176
Mechanical Factors	184
SERVICE IN THE BRITTLE TEMPERATURE RANGE	189
REFERENCES	191

CHAPTER 10 FABRICATION DEFECTS	194
EXPERIENCE OF FABRICATOR	195
PIPE BENDING	195
Cold Bending	195
Hot Bending	195
Ovality	200
EXTRUDING OF OUTLETS	201
SWEDGING OF REDUCING ENDS	202
REFERENCES	204

CHAPTER 11 END PREPARATION AND FIT-UP FOR WELDING PREPARATION OF WELDING BEVELS	204
WELD JOINT FIT-UP	207
Double Bevel Joints	207
Pipe Joints With Backing Rings	208
Pipe Joints Without Backing Rings	210
Socket Welds	210
REFERENCES	213

CHAPTER 12 WELD DEFECTS	214
CORRELATION OF TEST RESULTS TO SERVICE BEHAVIOR	214
DEFECT CLASSIFICATIONS	217
ARC STRIKES	217
BACKING RINGS	222
BURN THROUGH	223
CENTERLINE CREVICE	226
CRACKS	226
Hot Cracking	229
Effect of Welding Process	233
Cold Cracking	233
Microfissuring	234
Base-Metal Cracking	235
CRATER PITS	238
HI-LOW	239
INCOMPLETE FUSION	241
LACK OF PENETRATION	242
OVERLAP	245
OXIDATION	245
POROSITY	247
SINK OR CONCAVITY	250
SLAG INCLUSIONS	253
SLUGGING	256
TUNGSTEN INCLUSIONS	256
UNDERCUT	259
WAGON TRACKS	260
WELD REINFORCEMENT	263
WELD SPATTER	266
WELD STRESSES *(not indexed)*	266a
REFERENCES	267
CHAPTER 13 EFFECTS OF WELDING ON DEFECTS AND FAILURES COMMONLY USED WELDING PROCESSES	272
WELDING PROCESS CONSIDERATIONS	273
Submerged-arc Welding	273
Inert-Gas Tungsten-arc Welding	273
Gas Shielded Consumable Metal-arc Welding	277
Flash and Resistance Welding	279
WELDING PROCEDURE CONSIDERATIONS	281
Welding Filler Metal Controls	281

Inert-Gas Tungsten-arc Welding	284
Submerged-arc Welding	286
Columbium-bearing Stainless Steels	287
Dissimilar Metal Joints	289
Titanium Vessels and Pipe	289
FAILURE CONDITIONS	290
Alignment Welds	290
Lack of Penetration	292
Arc Strikes	295
Hardness Variations	296
Base Metal Inclusions	297
WELDING OF EQUIPMENT IN SERVICE	297
REPAIR OF WELD DEFECT	298
REPAIR OF FAILURES	299
REFERENCES	299

CHAPTER 14 HEAT TREATMENT	301
PURPOSES	301
PREHEAT TREATMENT	302
POSTHEAT TREATMENTS	305
Uniform Hardness	305
Close Temperature Control	306
Solution Heat Treatment of Corrosion-Resisting Stainless Steels	307
Solution Heat Treatment of Superheater Tubing	308
Quenching and Tempering Heat Treatments	309
Indiscriminate Elimination of Heat Treatment	309
Nonuniform and Localized Heating	312
Excessive Heating	316
FLAME STRAIGHTENING	316
PROPER HEAT TREATMENT	319
REFERENCES	323

CHAPTER 15 CLEANING	325
PICKLING	325
Shot and Sand Blasting	327
TURBINING	328
GRINDING OF WELDS	329
SHOT CLEANING OF BOILERS	330
IMPROPER CLEANING FOR REPAIR WELDING	331
REFERENCES	332

CHAPTER 16 SERVICE FAILURES—STRESS AND FATIGUE CAUSES OF FAILURE — 333

- EXCESSIVE STRESSES — 335
- OVERPRESSURE — 335
- EXTERNAL LOADING — 336
- MECHANICAL FATIGUE AND SHOCK — 336
  - Definitions — 336
- MECHANICAL FATIGUE — 337
- MECHANICAL SHOCK — 339
- THERMAL FATIGUE AND SHOCK — 340
  - Definitions — 340
  - Intermittent Operation — 341
  - Operations Involving Temperature Cycling — 346
  - Thermal Fatigue In Process Equipment — 350
  - Notch Effects — 351
- OVERHEATING — 352
- HYDROGEN EMBRITTLEMENT AND DAMAGE — 355
  - Definition — 355
  - Failure Progression — 355
  - Hydrogen Diffusion Into Steel — 358
  - Methane Formation — 359
  - Effects of Time and Temperature — 360
  - Effects of Heat Treatment and Alloying — 363
- HYDROGEN BLISTERING — 365
- OXIDATION INSIDE CRACKS — 365
- EXCESSIVE SERVICE CONDITIONS OR INADEQUATE DESIGNS — 366
- REFERENCES — 367

CHAPTER 17 SERVICE FAILURES—CORROSION — 370

- TYPES OF CORROSION — 370
  - General Corrosion — 372
  - Galvanic Corrosion — 373
  - Minor Dissimilarities in Composition — 374
- INTERGRANULAR CORROSION — 377
- KNIFE-LINE CORROSION — 377
- PITTING CORROSION — 379
  - Carbon Steel — 379
  - Aluminum — 382
  - Crater-Type Corrosion — 382
  - Stainless Steels — 383

CREVICE CORROSION	385
STRESS-CORROSION CRACKING	387
Nature of Cracking	388
Caustic Embrittlement	394
Oxidation Stress Corrosion	398
CORROSION FATIGUE	398
Elevated Temperature Service	403
Remedies	406
STRESS-ENHANCED CORROSION	406
EROSION	409
EXFOLIATION	414
SOIL CORROSION	415
CORROSION CAUSED BY INSULATION	416
SCALING AND OXIDATION	416
ANALYSIS OF FAILURES	417
REFERENCES	418
CHAPTER 18 MEANINGFUL INTERPRETATION OF INSPECTION RESULTS *(not indexed)*	421
INDEX	439

*Defects and Failures in
Pressure Vessels and Piping*

———chapter 1

INTRODUCTION

DEFINITIONS

Perfect pressure vessels and piping materials do not exist.

A defect is normally defined as *the lack or absence of something essential for completeness or perfection.* Defects may range from atom-sized dislocations in the metal, which cannot be observed with the highest powered microscopes, to major metal discontinuities that are visible to the eye.

Failure of a material is defined as *omission to perform or non-performance—often involving deterioration or decay.* Many apparent defects that are readily visible may not reduce the service life of the pressure vessel or piping component. On the other hand, many visible surface defects and invisible subsurface defects may be harmful and can result in service failures.

Failures may also result in nearly perfect materials due to external causes such as poor design, corrosion, or improper operation of the equipment in which the material is used.

Under one set of service and operating conditions, a particular defect may be inconsequential and extremely unlikely to cause or contribute to a service failure. Under another set of conditions, the same defect can be extremely hazardous and cause an instant rupture leading to explosions, fire and other damage and endangering the life and health of people.

Defects are normally considered to represent separations in the normal metal structure. The word "defect" is broadly used. Various other terms have also been used to describe defect conditions. They are inclusions, nonhomogeneities, flaws, structural discontinuities, non-perfections, etc. However, in a broader sense, the word "defect" may also describe *dissimilarities* in metallurgical structure or *design weaknesses*.

LABORATORY TEST EVALUATION OF DEFECTS

It must be recognized that each specific destructive or nondestructive test will evaluate a *particular* material in comparison with another material tested under the *same* conditions. The interpretation of test results and their correlation with actual service characteristics at best are only approximate. Quite frequently, the results have little or no relation to the actual service performance of the material or weldment.

Almost any defect can be caused to produce a crack and be subsequently propagated into failure by one kind of a severe test or another. The same defect, however, may be completely inconsequential when located in the base metal or in a weld joint in a pressure vessel, tank, or piping system subjected to service conditions also considered extremely severe. For example, in tension fatigue tests, the weld ripple on a cover pass can be readily proved to produce cracking and cause failure. However, similar or far more severe weld ripples have never caused a failure in hundreds of thousands of critical pipe and boiler tube welds involved in the major high-temperature high-pressure steam power plant piping systems and boilers now in operation. Ordinary weld ripples should be viewed with no greater concern than surface laps or slivers commonly found and accepted on plate and pipe materials.

There are indications that crack initiation in laboratory fatigue tests is enhanced by stress-oxidation. For example, the exclusion of air by the application of epoxy coatings to steel fatigue test specimens involving notched base plates and transverse butt welds increased fatigue life by a factor of 3.[1] Fatigue testing thus could be considered as a stress-corrosion mechanism. In service protective coatings, scale or environmental conditions may exclude oxygen from the metal surface subjected to fatigue.

In a recent analysis of test data on which various crack propagation laws have been based, it was concluded that the practice of using data from single test specimens is not a sensitive evaluation of the validity of a crack propagation law.[2] Data chosen at random from single specimens led to an apparent agreement of several contradictory laws. The authors believed that the conclusions, often hastily drawn, needed to be reexamined before any given crack-propagation law is accepted as valid.[2]

On the other hand, tests properly selected and carefully planned can provide useful results. When realistically interpreted, the test results may provide a better understanding of the causes leading to a specific failure, and aid in the development of changes in the specifications or in the operational procedures reducing future failure hazard of a specific defect condition.

CODE RECOGNITION OF DEFECTS

The major construction codes applicable to pressure vessels, tanks, and piping generally recognize that perfect materials do not exist, and make allowance for the presence of defects by establishing limitations on defect type, size, location, and distribution.

Unfortunately, the effects of many defects on the service life of pressure vessels and piping are largely misunderstood by engineers. Great emphasis and tight limitations are applied to some defects, whereas others are ignored even though they may represent a greater hazard to the service life of the equipment.

Basing requirements in specifications only on the result of small section laboratory tests may lead to unrealistic conclusions, and has been a cause in a significant number of major failures.

INSPECTION FOR DEFECTS

The testing and inspection techniques developed to evaluate the soundness of fabricated structures have contributed significantly to the satisfactory utilization of engineering materials under increasingly severe service conditions. New and refined testing and inspection facilities can also give engineers greater confidence in the safety of the systems for which they are responsible. This is a matter of concern to plant managements, since service failures can be extremely hazardous and costly with the increasingly severe operating pressures and temperatures involved in industrial processes. However, the confidence of managements and engineers in the results of inspection is justified only if the testing methods are carefully selected and if properly qualified persons perform the testing.

Unfortunately, a false sense of security can be engendered when a fabricated pressure vessel, tank, or piping system has been thoroughly inspected and defects have not been detected. Potentially dangerous conditions can defy some or all of the standard testing equipment available and the techniques used. Such hazardous conditions exist now in pressure vessels, tanks, and piping components. They may not be detected until a sudden rupture occurs.

BRITTLE AND DUCTILE FAILURES

Among the various terms frequently used in describing potential failures are brittleness, toughness, and ductility.

Brittleness is the property of a material that leads to crack propagation without appreciable plastic deformation.

Toughness and ductility are essentially antonyms of brittleness. *Toughness* describes the ability of a material to absorb energy and deform plastically before fracturing. By representing the work required to cause fracture, toughness is a measure of both strength and ductility.

Ductility describes the ability of a material to deform plastically without fracturing. It is generally measured in tensile tests by uniform elongation before necking, or by the reduction in the area between onset of necking and fracture.

Sudden ruptures of pressure vessels and piping generally are viewed with the greatest concern. These are normally associated with the notch-brittle behavior of certain mild and low-alloy steels. Many brittle failures have resulted in substantial damage to other equipment and structures, and sometimes have caused severe injury and loss of life. These notch-brittle failures are normally associated with service at atmospheric or subfreezing temperatures.

Introduction 5

Sudden ruptures, however, have also occurred at elevated temperatures. In several instances, for example, where piping carried steam, severe damage and loss of life were also caused. In these brittle failures, the metal itself has been severely embrittled internally by metallurgical changes. Stress, to a far greater extent than temperatures, has triggered the sudden rupture.

Fortunately, by far the majority of failures occur in a ductile manner. They involve either the gradual propagation of cracks or corrosion across the wall thickness of the pressure vessel, tank, or pipe. By bulging or leaking ductile failures give a warning to the operating personnel. Without major damage to the equipment, repair or replacement, or temporary repair followed by subsequent replacement of the damaged section is then possible.

These ductile failures have occasionally resulted in severe damage and loss of life when the pressure vessel, tank, or pipe contained inflammable liquids or gases. For example, in 1955, in an oil refinery near Philadelphia, a leak developed in the rundown line from the crude units to the receiving house. Flow from the crude units was halted immediately. However, gas spread to the receiving house basement, and was ignited from an unknown source. Management, maintenance and operating personnel working to repair the leak were burned. Although the fire was controlled in 45 minutes, and property damage was negligible, 18 persons were injured, 6 critically.[3]

In August 1955, losses in excess of $1 million resulted from the explosion of a new fluid catalytic hydroformer near Chicago. Flames quickly spread to surrounding gas and oil tanks. Debris crashed into nearby houses and littered the streets. One fatality occurred, and injury resulted to 36 persons. It was believed that the blast failure resulted from a material failure, from abnormal stresses or pressure, or from the introduction of water into the unit's high temperature section.[4]

In 1951, a thermometer well joint failed in the low pressure return oil line at the turbine end of a 30,000 KW turbo-generator near Columbus, Ohio. The resulting leak allowed oil to escape under 15 lb pressure, and to ignite at surfaces heated to 900°F by superheated steam. Oil discharged into the fire for 45 minutes. When the fire had burned itself out, unprotected steel roof trusses 50 ft overhead had collapsed and about 6000 sq ft of the concrete roof of the noncombustible building had fallen in. The damage resulting from the small leak amounted to $1½ million.[5]

A pipe failure has been credited as the most likely cause of the sinking of the nuclear submarine Thresher in 1963 in which 129 persons lost their lives.

In nuclear reactor primary piping systems, leaks caused by gradual cracking through a ductile metal may result in the loss of heavy water moderators controlling reactor heat output. Other potentially hazardous consequences of leaks may be the flooding of the circulating pumps and the failure of the reactor cooling system.

Loss of steam in boiling water reactors may decrease heat removal from the reactor and cause positive reactivity, leading to increased reactor output.

Although brittle failures are often considered the most catastrophic, it is important to recognize that cracking through ductile metal may also have disastrous consequences.

REFERENCES

1. Gilde, W., "Increasing the Fatigue Life of Butt-Welded Joints," *British Welding J.*, **7**, 208–211 (1960).
2. Paris, P. and Erdogan, F., "A Critical Analysis of Crack Propagation Laws," *Trans. ASME*, **85**, Series D, 528–534 (1963).
3. *Oil and Gas J.*, **53**, 87 (Oct. 17, 1955).
4. *Chem. Eng. News*, **33**, 3664 (1955).
5. National Fire Protection Association, Fire Record Bulletin FR 55–1 (1955).

―――chapter 2

CAUSES OF SERVICE FAILURES

CLASSIFICATIONS

Although numerous conditions may cause or lead to service failures, the responsibility for a failure can generally be assigned to one of five classifications:

(1) Design (structural, design notches, joint location, or welding end configuration)
(2) Materials (selection and handling of base and welding materials)
(3) Base Metal Defects (introduced during manufacture and shaping of plate or piping components—pipe, cast valve, cast or forged fitting, etc.)
(4) Fabrication (fabrication, welding, heat treatment, or cleaning of pressure vessels or piping during shop fabrication or field erection)
(5) Service (excessively severe service conditions).

In some instances, the responsibility can be related to a combination of several of these classifications. For example, a pipe weld containing root defects, such as lack of penetration, may not fail during service until thermal or mechanical fatigue of sufficient severity causes cracking and crack propagation of the existing defect across the wall thickness.

If complete weld root penetration was originally specified, then the failure should be classified as follows:

Primary Classification: Fabrication
    Condition: Improper welding, lack of penetration in root of weld
Secondary Classification: Service
    Condition: Thermal fatigue

On the other hand, if the weld quality was not specified at the time the weld was made, and some incomplete weld root penetration was considered acceptable since thermal fatigue service conditions were not anticipated, then the failure should be classified as follows:

Primary Classification: Service—excessively severe
    Condition: Thermal fatigue
Secondary Classification: Fabrication
    Condition: Lack of penetration in root of weld, considered acceptable under applicable specifications.

This illustrates the importance of familiarity with the service requirements when preparing job specifications and fabrication procedures.

FAILURE MECHANISM

A failure occurs by cracking or corrosion or sometimes a combination of both. The majority of failures occur gradually. Upon traversing the wall thickness of the pressure vessel, tank, or piping component, a surface opening develops through which some of the steam, liquid, or gas carried in the system may leak. This, in effect, produces a warning of the potentially hazardous condition. Very rare are failures where partial cracking across the wall occurs, followed by a sudden pipe rupture, or where a sudden rupture occurs which is not preceded by detectable prior cracking. These sudden-rupture type failures are viewed with extreme concern as they may result in injury of personnel or loss of life, and may become extremely costly to the plant. Most of these failures are related to the notch-brittle behavior of certain steels.

Where failures occur by cracking, three stages are defined. They are:

(1) Initiation
(2) Growth
(3) Propagation

Crack Initiation

The *initiation* of a failure involves the conditions leading to or causing the initial cracking. Such cracking may be of submicroscopic or microscopic size, or may even be visible to the eye. Three conditions are involved in crack initiation:

(a) Stuctural defects and nonhomogeneities present in the original components.
(b) Metallurgical defects and nonhomogeneities present in the original components.
(c) Defects introduced during service in structurally and metallurgically sound materials.

Generally, only the first condition is searched for by means of nondestructive inspection techniques. It must be recognized that perfect plate and piping materials and welds are not made commercially.

Structural defects may range from atom-size dislocations in the metal, which cannot even be observed under the highest power microscopes, to major metal discontinuities visible to the eye. Many gross defects, such as base-metal inclusions, quantities of weld porosity, or slag discernible on radiographic films may not reduce the service life of the pressure vessel, tank or piping component. However, many other visible surface and invisible subsurface defects can result in service failures.

Metallurgical defects represent differences in properties of sufficient magnitude to become a potential cause of failure. Such conditions can occur within the same metal, for example, where localized heat treatment produces an area of high hardness adjacent to an area of lower hardness. Metallurgical notches also can represent a higher hardness level in the heat-affected zone adjacent to base metal or weld metal, or may be due to differences in properties between a wrought base metal and a weld deposit, which can be considered to represent essentially a casting. Localized cold work or residual stresses also are included. More apparent metallurgical notches occur between dissimilar metal combinations, as clad materials, weld overlays, dissimilar metal welds, etc.

Defects introduced during service involve cracks, corrosion pits, grooves, or general corrosion occurring at locations where the material itself was structurally and metallurgically sound in the originally erected condition.

Crack Growth

The initiation and extension of the crack to a critical size is defined as the *initial growth* stage. The extent depends on factors such as the mechanical and metallurgical properties of the material, the nature of the initial defect, the size and thickness of the vessel or pipe and the stress level caused by residual and external stresses and loads. In inherently brittle materials, the growth stage may be almost infinitesimally small, whereas in a ductile material, the crack may extend into the wall thickness to a significant extent.

Crack Propagation

The propagation stage involves the *unstable stage* beyond the growth stage when the crack extension becomes more rapid. Such propagation may be continuous or intermittent. It may be extremely rapid, as occurs often in brittle materials. In a ductile material, such crack propagation is generally slow, and may take days, months, years, or decades before it traverses the wall thickness of the pressure vessel or pipe. Sometimes crack propagation in a ductile material may occur slowly in the earlier stages, and then continue very rapidly in the final stages. Thus, such failures may have the appearance of brittle failures, even though the mode of fracture is primarily of the shear type. In these instances, a severe mechanical or thermal shock or external load often triggers the final sudden rupture.

In new installations where long service histories are not available, the correlation of potential failures in some critical service environments is best approximated by an examination of actual failures in pressure vessels and piping in other types of commercial plants rather than by laboratory test data. By determining the factors responsible for initiating and propagating the cracking, corrosion or erosion conditions leading to failures, allowance may then be made in the design and in the specifications covering materials, fabrication and inspection for either the prevention of such failures, or for the installation of safety devices minimizing the dangers inherent in the occurrence of ruptures or other hazardous explosions.

chapter 3

DESIGN CONSIDERATIONS

Failures associated primarily with design may be the result of improper structural design, often involving insufficient flexibility in the piping system. Involved also are failures which start from design notches, such as sharp corners, reinforcements, or attachments to the vessel, tank or piping component which cause severe localized stresses or restraints. Finally, these types of failures may start in welds where the weld-joint design or machined welding end preparations make deposition of a sound weld difficult.

In cases where design is conducive to failure, thermal fatigue, mechanical fatigue, or corrosion are often the causative factors in initiating cracks and propagating them across the wall thickness so that failure results.

FLEXIBILITY

Failures resulting from insufficient flexibility in vessels, headers, tanks, or piping systems represent fairly common conditions of service failures. They are particularly apt to occur in piping systems subjected to thermal or mechanical fatigue.

Figure 3-1 illustrates a steam lead that failed as a result of insufficient flexibility. During periodic shutdowns of the boiler, thermal contraction occurred in the piping system. This was followed by expansion of the metal during operating periods involving steam

Design Considerations 13

Fig. 3-1. Cracking failure in steam piping caused by insufficient flexibility.

flow through the piping system. With insufficient flexibility in the system to absorb the stresses over the years, cracking was initiated and gradually spread across the wall thickness. Such cracking frequently occurs near anchor points or major branch connections, which act as a major restraint on the free contraction and expansion of the system.

Pipe headers are also susceptible to these failures, particularly where supports restrain movement that results from expansion of the major steam leads connecting into the header. This is illustrated in Fig. 3-2. The bending of the pipe over these restraints may lead to cracking and eventual failure. Even free-floating headers may fail

14   *Defects and Failures in Pressure Vessels and Piping*

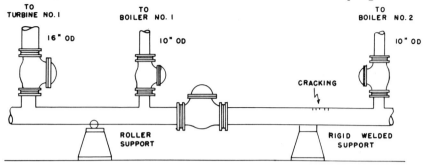

Fig. 3-2. Cracking in pipe header section near restraint over which bending of pipe occurs.

when pipe leads joined to the header do not have sufficient flexibility and tend to pull the header in different directions.

Thermal fatigue failures generally occur circumferentially. The majority of these cracks originate on the inside of the vessel, tank, or pipe, especially if the wall thickness is $\frac{1}{2}$ in. or more. Frequently, thermal fatigue initiates a family of parallel cracks, as shown in Fig. 3-3. In such cases, a weld repair of the first crack to traverse the wall thickness may soon be followed by a second leak and another plant shutdown to repair a new crack in the same vicinity.

Figure 3-4 illustrates examples of good and poor design of instrumentation and steam sampling lines. Excessive rigidity and high

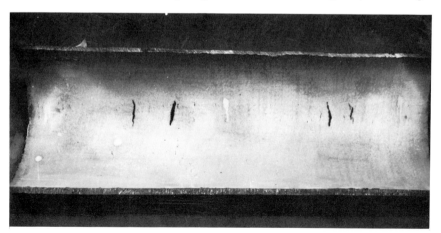

Fig. 3-3. Typical parallel cracking in pipe section near rigid restraint.

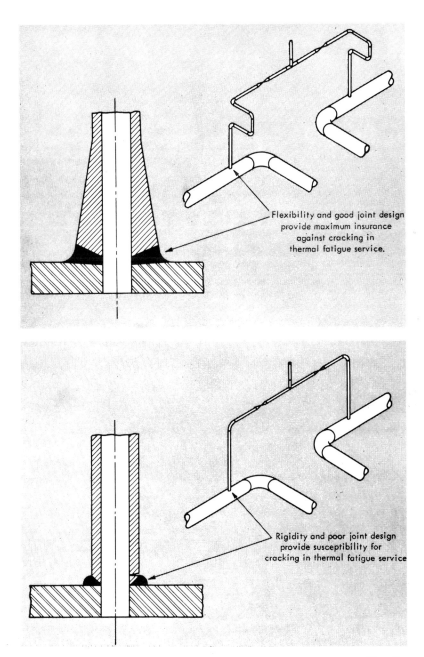

Fig. 3-4. Proper and poor flexibility and joint design in steam sampling line.

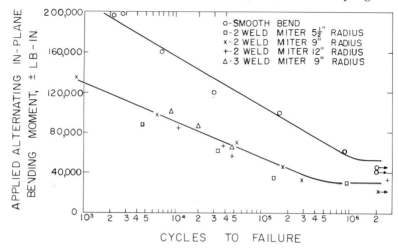

Fig. 3-5. Test results comparing effects of applied alternating in-plane bending moment against number of cycles to failure in miters and bends.[1]

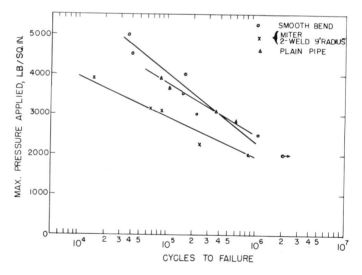

Fig. 3-6. Test results comparing effects of maximum pulsating pressure against number of cycles to failure in straight pipe, a bend, and a miter.[1]

stress concentration at the nozzle weld make these lines highly susceptible to cracking.

Bends and Miters

Generally, the designer also makes the decision on the selection of elbows, pipe bends, or miters. Alternating in-plane bending fatigue and pulsating pressure fatigue tests on 6-in. diameter by 0.278-in. wall pipe have shown a significant reduction in strength for miters as illustrated in Figs. 3-5 and 3-6. The flexibility factors under in-plane bending for miters tested were approximately 80 to 85% of those predicted for equivalent smooth pipe bends.[1]

Tanks

In tanks subject to mechanical fatigue, insufficient rigidity has also led to service failures. Failures in welded water tanks on locomotive tenders occurred repeatedly after 9300 to 20,500 miles of operation until the rigidity of the tanks was increased sufficiently with gussets and profile stiffeners at the points exposed to the greatest stresses. Subsequent service was satisfactory for over 620,000 miles of service.[2]

DESIGN NOTCHES

Sharp Corners

Good design practice should recognize that when a pressure vessel or piping system is to be subjected in service to thermal or mechanical fatigue, cracking may start at surface or internal notches. Sharp corners should be especially avoided where fatigue is involved.

Figure 3-7 illustrates cracking on the inside of a turbine casing. The cracking started at sharp corners. The repair procedure involved grinding out of the cracks. The area was preheated and rewelded, and subsequently ground to provide uniformly smooth contours. Finally, the shell was stress relieved. Cracking has not occurred during subsequent service.

A sharp design notch condition involving a welded joint in a main steam piping system which caused cracking is illustrated in Fig. 3-8. After several months of operation at 950°F, the joint had cracked through the wall for almost 180 degrees of the pipe circumference.[3]

18 *Defects and Failures in Pressure Vessels and Piping*

Fig. 3-7. Ground-out cracks at inside of turbine casing originating at sharp design corners.

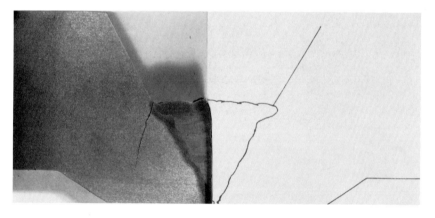

Fig. 3-8. Crack in a welded pipe joint due to notch concentration. Left side is photograph of section containing crack as removed from pipe; right side is sketch of remaining portion of weld and part of steam line.[3]

Sharp corners have also led to a significant number of brittle failures, and should be particularly avoided in steel vessels and tanks subject to service at temperatures below the transition temperature of the steel used (see Chapter 9).

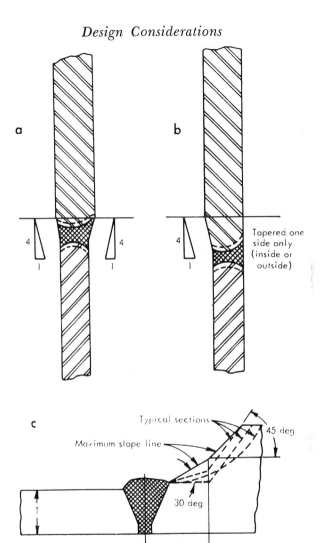

Fig. 3-9. Recommended weld end preparation for pipe, valves, and fittings of unequal thickness.

Changes in Wall Thickness

A number of failures have occurred in machined, cast, and welded components involving both thick and thin sections where cracking has started from undesirable design notches.

To minimize the possibility of these notch types of failures, the

Fig. 3-10. Poor design of flange connection to engine exhaust tubing subject to mechanical fatigue.

design should provide for gradual changes in section thickness and for rounded rather than sharp corners. In this connection, many Codes recommend a 4 to 1 taper in joints involving materials of different wall thicknesses. This is illustrated in Fig. 3-9.

Joints between slip-on flanges and pipes represent design conditions susceptible to failure when the service involves mechanical or thermal fatigue. A common failure is shown in Fig. 3-10 which involved a flange connection joining exhaust tubing to a car or truck engine.

A similar design produced failure in a flanged joint of the low-pressure stage of an air compressor.[4] Fracture occurred after 350 hours of service along the weld between the slip-on flange around a 7-in. diameter elbow (Fig. 3-11). Replacement by a welding neck flange resulted in trouble-free service. Improper design of the original connection was considered the primary cause of failure. The abrupt joint between the rather flexible elbow section and the slip-on flange produced a maximum stress concentration. This was reduced

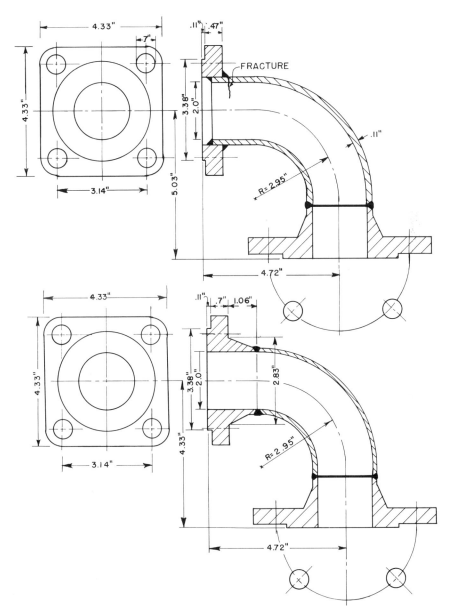

Fig. 3-11. Sketch illustrating design responsible for fracture in flanged elbow of air compressor, and improved design of replacement section.[4]

Fig. 3-12. Fatigue failure in socket weld involving thermometer well.

Design Considerations 23

significantly by the more gradual change in thickness of the welding neck flange.

Outlets

The proper design of outlets is quite controversial. Satisfactory outlet designs for one application may not be suitable for another involving a more severe service condition.

In fact, service failures have occurred in almost every type of outlet design. Tests may not always provide good correlation to service performance, as they may not make allowance for all environmental factors involved in the actual service.

Figure 3-12 illustrates a common failure in a socket weld joining a thermometer well to a steam pipe. Thermal fatigue has been a contributory factor to the failure caused by the inherently notch-sensitive design.

A mechanical fatigue failure in another nozzle weld is illustrated in Fig. 3-13. The joint design was too rigid for the severe mechanical fatigue conditions involved.

Fig. 3-13. Mechanical fatigue failure in nozzle weld providing excessive rigidity.

24 *Defects and Failures in Pressure Vessels and Piping*

Fig. 3-14. Cracking adjacent to weld between reinforcing ring and pipe header. Cross section illustrates path of crack propagation.

Design Considerations

A number of failures have also occurred in nozzle welds strengthened by reinforcing rings, particularly where severe thermal fatigue is involved.

Figure 3-14 illustrates cracking adjacent to the weld between the reinforcing ring and pipe header carrying steam in a power plant. The air space between the reinforcing ring and pipe wall contributed to temperature differences between the ring and pipe material. The differential expansion between these materials resulting from thermal fatigue produced the cracking.

Cracking on the inside of a clad stainless steel vessel is shown in Fig. 3-15. The wall thickness was approximately 1 in. Cracking occurred underneath the welds joining a heavy reinforcing ring around

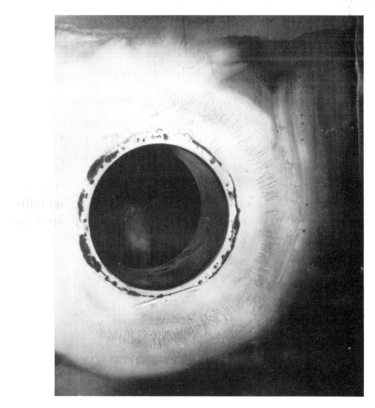

Fig. 3-15. Fatigue cracking on inside of clad stainless steel vessel underneath welds between reinforcing ring around nozzle on vessel wall.

nozzle outlets to the vessel wall. The service involved thermal fatigue cycling between approximately 150 and 500°F.

Stress-concentration Factors

In the design of nozzles, the presence of stress concentrations can have an important bearing on the service life.

The *stress-concentration factor* is normally defined as the ratio of the maximum stress to the nominal hoop stress.

Tests with different nozzle designs at room temperature employing strain-gage determinations on cylindrical model vessels generally show stress-concentration factors between 1.5 to 2.8 for different types of reinforcing plates.[5] These values are substantially lower than those obtained in tests on reinforced nozzles.

In fact, detailed room temperature fatigue tests on 20-in. diameter cylindrical vessels containing various non-reinforced and reinforced nozzles have indicated stress-concentration factors as high as 3.5.[6] Comparison of these results with data obtained from plain plate fatigue tests showed that up to about 1,000,000 cycles, the vessel failures were in fairly good agreement with the results of plain plate failure

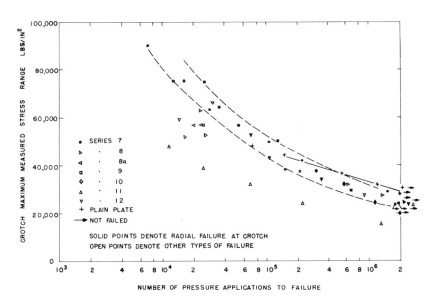

Fig. 3-16. Correlation of peak crotch stress in several designs of nozzles with plain plate fatigue strength.[6]

fatigue tests. This is illustrated in Fig. 3-16. In the vessel failures, the stress was considered in terms of the stress-concentration factor multiplied by the nominal hoop stress. Thus, at room temperature, the stress-concentration factor is considered to provide a relation to the vessel behavior. This particular test program [6] also illustrated that the unreinforced nozzles did not result in an appreciable reduction in fatigue life as over 100,000 applications of pressure to 25% of the ultimate tensile strength were required before leakage occurred. Moreover, reinforcement restricted to either inner surfaces or outer surfaces of the vessel did not have a significant effect on fatigue performance. However, there was an improvement when the reinforcement was equally spaced between inside and outside vessel surfaces, (Fig. 3-17, curve 12).

Other mechanical fatigue tests [7] conducted on vessels with the different large openings shown in Fig. 3-18 also provided generally good

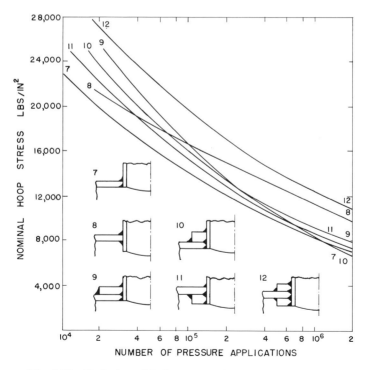

Fig. 3-17. Variation of fatigue strength with nozzle design.[6]

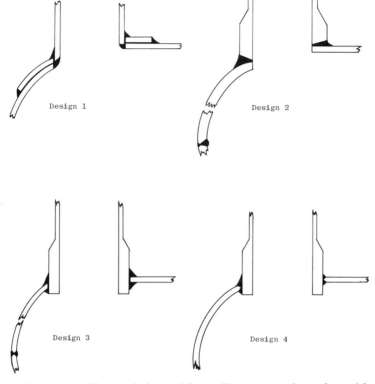

Fig. 3-18. Different designs of large diameter nozzle outlets with 8.9-in. ID and ½-in. wall.[7]

correlation between stress-concentration factors and fatigue life. The reduction in fatigue strength involving up to 100,000 cycles caused by the different openings was estimated to be:

33% for design 1 (max. stress-concentration factors ID − 2.0, OD − 1.9)

21% for design 2 (max. stress-concentration factors ID − 2.5, OD − 1.7)

21% for design 3 with defective internal fillet

12% for design 3 with perfect fillet weld (max. stress concentration factors ID − 1.5, OD − 1.5)

7½% for design 3 with ground weld

6½% for design 4 with ground weld

CODE KEY

SHOP CODE

1 2-2148	2 2-2149	3 2-2146
4 2-2420	5 2-2432	6 2-2159
7 H3300	8 _____	9 _____

WELD SPECIFICATION CODE

1 BAC 5940	5 BAC 5948	9 ASME. SEC. III
2 BAC 5975	6 AWS D1.1	10 ASME. SEC. VIII
3 BAC 5945	7 NSS 8002	11 D250-10046-1
4 BAC 5912	8 NSS 8003	12 _____

W/O

1 WELDER (BAC 5945)
2 OPERATOR (BAC 5945)
3 WELDER-CLASS A
4 OPERATOR-CLASS A
5 OPERATOR-PROD.
9 _____

NDT

1 PENETRANT
2 MAG PARTICLE
3 ULTRASONIC
4 X-RAY

WELD PROCESS CODE

1 SMAW ✓	4 SAW ✓	7 EGW ✓
2 GMAW ✓	5 FCAW ✓	8 PAW
3 GTAW ✓	6 ESW ✓	9 EBW
		0 OTHER

DEFECT CODE

1 UNAUTHORIZED WELDING
2 ELECTRODE/FILLER METAL
3 CURRENT/POLARITY
4 PREHEAT/INTERPASS TEMP
5 SEQUENCING IMPROPER
6 TRAVEL SPEED
7 WELDER UNQUALIFIED
8 CRACKS
9 POROSITY/PIN HOLES
10 UNDERCUT EXCESSIVE
11 LACK OF FUSION
12 LACK OF PENETRATION
13 OVERLAP
14 CONTOUR
15 BURN THRU/DROP THRU
16 WELD PROCESS
17 WELD MISSING/UNDERSIZE
18 OVERFILL
19 SLAG/EXCESS SPLATTER
20 WELD STAMP MISSING
21 INCLUSIONS/VOIDS
22 _____
23 _____

PROGRAM

1 ALCM
2 B-1 RDT&E
3 E-3A
4 E-4
5 MINUTEMAN
6 ROLAND
7 RSLP
8 SRAM
9 TRIDENT
10 YC-14
11 JETFOIL
12 NUCLEAR POWER
13 COMM NUCLR PROD.
14 COMM MECH MFG.
15 B-1 PROD.
16 _____
17 _____

REPORT CODE

1 UER   2 P/U   3 AS MARKED

WELD DATA INPUT

5218 | 30 | 31 | 32 | 33 |

PART NUMBER																	QTY		SHOP	SPEC	PROC	W/O	NDT	PROG		SKD	FLAG
1	2	3	4	5	6	7	8	9	10	11	12	13	14	15	16	17	18	19	20	21	22	23	24	25	26	27	28	29

1ST WELD STMP			INCHES OF WELD					DEFECT INCHES			DEFECT CODE		DEFECT INCHES			DEFECT CODE	
34	35	36	37	38	39	40	41	42	43	44	45	46	47	48	49	50	51	52	53

2ND STMP			INCHES OF WELD					DEFECT INCHES			DEFECT CODE		DEFECT INCHES			DEFECT CODE	
54	55	56	57	58	59	60	61	62	63	64	65	66	67	68	69	70	71	72	73

REPORT DATE — MO 74 75 | DAY 76 77 | Y 78 | INSP STMP

REPORT CODE 79 | 80

SHOP STMP | INSP STMP | DATE

REMARKS

L 30-035

Design Considerations

Other fatigue tests, particularly involving low-cycle fatigue conditions have also confirmed that peak stresses caused by different nozzle designs correlate with fatigue behavior.[8,9]

Fatigue tests on oblique nozzles on a vessel 66 ft. 9 in. in diameter by 4 in. thick with set-through vertical nozzles 12 in. in diameter welded with partial penetration butt welds showed that the maximum stress concentrations occurred in the shell at the end of the major axis of the nozzle of greater obliquity.[10] The highest stress-concentration factor determined was 2.5 compared with 1.8 for a similar position at a nozzle of small obliquity.

Nozzles at Elevated Temperatures

At elevated temperatures, the relations are considerably more complex. In certain types of reinforcements, the stress-concentration factor can be increased substantially by thermal fatigue and differences in temperature between shell of vessel or pipe and reinforcing ring. Moreover, where the service involves temperatures in the creep range, some relaxation may occur.

Where the service involves severe thermal fatigue cycles, the nozzle reinforcement designs shown in Fig. 3-19 may be advisable.[11]

Attachments and Supports

Where the service involves mechanical or thermal fatigue, the design of welded attachments should minimize notch conditions. The attachments and welds should be contoured uniformly and provide a smooth finish.

Figure 3-20 illustrates a poor and an improved design of reinforcing ribs on a structural pipe section. The mechanical fatigue service failure in Fig. 3-20A was avoided by gradual tapering of the ends of the ribs to the pipe surface and providing a smooth weld contour.

Special grinding or machining may be desirable as illustrated in Fig. 3-21. In the application illustrated, such a change increased the fatigue life by 70%.[12]

Where the design includes attachments for hangers or supports, heavy welds along the ends may act as sufficiently severe restraints to initiate cracks, again as a result of thermal or mechanical fatigue, or both. Figure 3-22 illustrates a thermal fatigue failure in a pipe bend in the main steam piping in a steam power plant. The piping

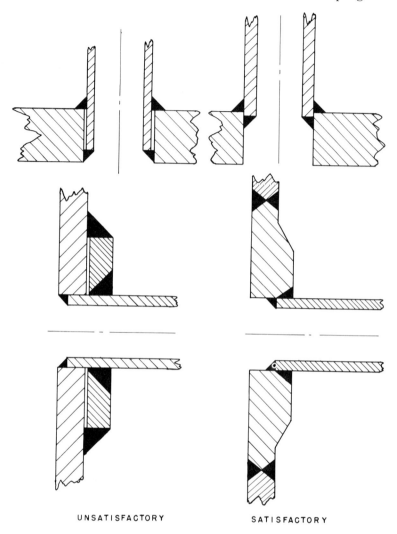

Fig. 3-19. Nozzle reinforcement design recommended for service involving thermal fatigue.[11]

material was a ½Mo steel 10-in. OD by 0.690-in. wall. Cracking started in the pipe adjacent to the fillet weld between the front end of the hanger lug and the pipe after 16 years of service at 890°F.

Similar conditions after 60,000 hours of operation led, in 1960, to a 20-in. long crack in a pressure-reducing pipe section of 0.3Mo steel changing in diameter from 19.7 in. to 9.85 in. Pipe supports were

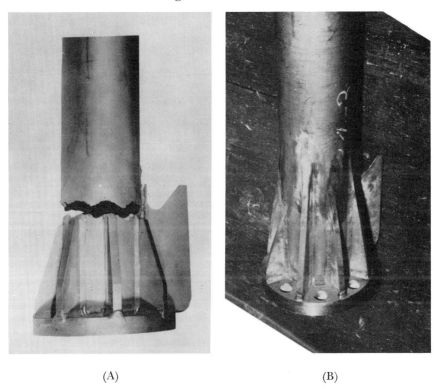

(A) (B)

Fig. 3-20. Examples of poor and improved design of reinforcing ribs in application involving severe mechanical fatigue of 5-in. Schedule 40 17-4 PH stainless steel pipe section.

welded with heavy beads on the outside. Preheating and post heat treatments were not applied to the attachment welds resulting in a hard martensitic structure. Mechanical fatigue was considered the major factor causing crack propagation. Elimination of the heavy attachment welds should have avoided this failure.[13]

The design of attachments and supports is also very important in chemical plant equipment subject to corrosion or erosion. A pressure vessel mounted on a railroad car and containing liquid gas at 120 psi exploded as a result of prior fatigue cracking which started from the pressure vessel supports.[14] The bottom was thrown a distance of over 300 feet. Although the pressure vessel was constructed out of a notch tough steel, the support material in which cracking started was made of a notch-brittle steel.

32 *Defects and Failures in Pressure Vessels and Piping*

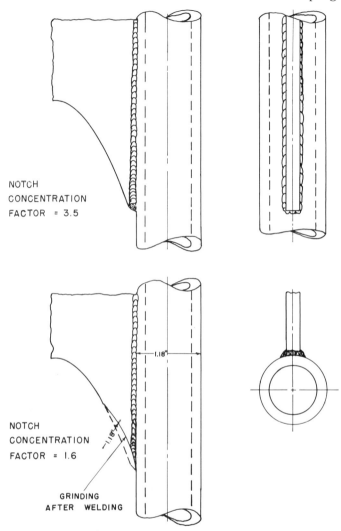

Fig. 3-21. Reduction of notch concentration factor by contouring after welding.[12]

Where gases or fluids are in motion, the placement of attachments or baffles may promote or reduce erosion.

Figure 3-23 illustrates failure produced by localized erosion in a Type 316 stainless steel heating coil placed around the inside wall in a 9-ft diameter reactor tank containing polyamide resins or urea-

Design Considerations 33

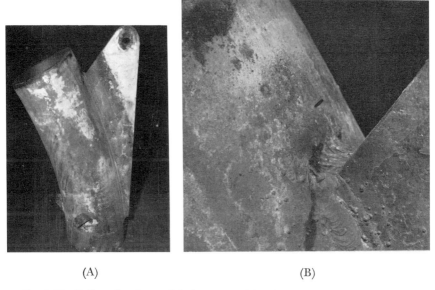

(A) (B)

Fig. 3-22. Failure by thermal fatigue at welded hanger attachment because of sharp change in weldment design. (A). Reduced to $\frac{1}{25}$. (B). Reduced to $\frac{1}{4}$.

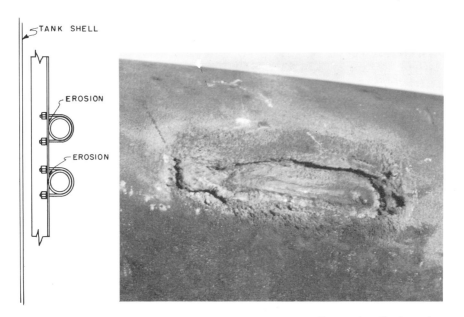

Fig. 3-23. Failure caused by U-shaped coil support leading to localized erosion due to turbulence.

34   *Defects and Failures in Pressure Vessels and Piping*

Fig. 3-24. Erosion of inlet to safety valve. It was avoided by the installation of baffles.[15]

formaldehyde resin solutions in water. The $2\frac{7}{8}$-in. diameter by 0.125-in. wall coil was supported by U-shaped clips $2\frac{7}{8}$-in. wide covering one-half of the coil circumference. The slurry was rotated rapidly, causing turbulence in the area where the tubing was in contact with the supporting angle. By changing the design of the supports or reinforcing the tube circumference with a uniformly contoured split ring, this localized erosive condition could have been avoided.

Sometimes it is desirable to install baffles in order to avoid turbulence. Figure 3-24 illustrates steam erosion (possibly with condensate) in a safety valve steam pipe leading to the high-pressure stage of the turbine. The steel was alloyed with $\frac{3}{4}$Cr–$\frac{1}{2}$Mo. The failure by erosion occurred after 20,000 hours at 985°F. Recurrence of the erosion was avoided in the replacement unit by the installation of baffles.[15]

WELD JOINT DESIGN

The design of the ends to be joined also has a significant bearing on the soundness of a finished joint. With few exceptions, pressure vessels, tanks, and piping systems are generally welded. They contain

Design Considerations

relatively few, if any, flanged or screwed connections. The proper design of welded joints is of major importance.

In specifying a particular joint design in pressure vessels or piping, the primary concern should be for the welder whose task it is to produce sound welds. Unfortunately, many design and welding engineers do not recognize this and show joint designs that are difficult to weld, not readily accessible, or prone to defect conditions. Of course, an important consideration is cost. The cost of a joint design must be weighed against its ability to minimize notch conditions or weld defects.

It should also be kept in mind that the different processes applied in welding may call for different joint designs.

The easiest and most economical welded joint consists of butting together two ends with double V or single V-bevel preparations, particularly when the bevels have been shaped by flame-cutting methods. Welding of the joint with covered electrodes by shielded metal-arc welding, by submerged-arc welding, or by other welding processes may leave lack-of-penetration areas at the root of the weld. The danger of such defects is particularly acute where welding has to be done from one side, as is generally the case in pipe welding where the inside is not accessible for back welding, repair or grinding. These root surface defects are extremely undesirable for services involving subzero temperatures, high temperatures or pressures, thermal or mechanical fatigue, or stress corrosion.

Backing Rings

Full penetration welds are required under the ASME Boiler and Pressure Vessel Code and under many other codes. In piping systems, and in some pressure vessels where the weld is not accessible from the inside, particularly where full penetration is considered essential, metal-arc welding joint designs frequently call for backing rings to back up the root of the joint. For pipe of relatively light wall thickness employed in normal service conditions flat backing rings are usually satisfactory.

In some noncritical service applications, ridge-type backing rings have been used since they facilitate fit-up and reduce the amount of weld metal required to fill the weld joint. The resulting weld, however, is inherently notch-sensitive and susceptible to cracking in

36   *Defects and Failures in Pressure Vessels and Piping*

service environments involving mechanical or thermal fatigue. An example of crack propagation through a butt weld in an 8-in. by 0.322-in. wall steel pipe carrying steam at 358°F is illustrated in Fig. 3-25.

Fig. 3-25. Cracking starting at notch caused by ridge type backing ring.

In the more critical piping systems where pipe of heavier wall thickness is used, proper fit-up of a flat backing ring may be difficult, and notch conditions susceptible to failure, as shown in Fig. 3-26, may result. For critical piping systems in power plants, such as main steam, hot reheat and boiler feed piping, machining of the pipe ends

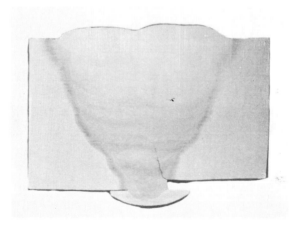

Fig. 3-26. Root crack at notch formed by improper fit-up of backing ring.

Design Considerations

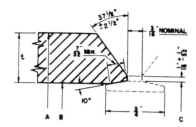

For Wall Thickness (t) 3/16" to 3/4" inclusive, and Tapered Internal Machining

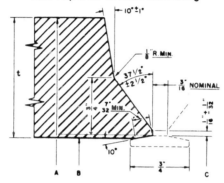

For Wall Thickness (t) Greater than 3/4" and Tapered Internal Machining

$$C = A - 0.031 - 1.75t - 0.010$$
$$= A - 0.041 - 1.75t$$

Where A = Outside Diameter in Inches
0.031 = Minus Tolerance on O.D. of Pipe to ASTM Specifications Listed Above
1.75 = Minimum Wall of 87½% of Nominal Wall, (permitted by ASTM Specifications listed above) Multiplied by Two to convert into terms of Diameter
t = Nominal Wall Thickness in Inches
0.010 = Plus Machining Tolerances on Bore Only

Fig. 3-27. End preparation for pipe butt welds for critical service applications employing taper machined backing rings.

and taper machined backing rings have provided very satisfactory joints, Fig. 3-27.

Elimination of Backing Rings

Special bevel preparations, such as the one shown in Fig. 3-28, represented an early attempt to eliminate backing rings. In the awkward positions of field welding, however, it is somewhat of a prob-

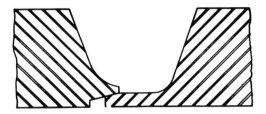

Fig. 3-28. Overlap type root preparation occasionally used in pipe butt joints to eliminate backing rings.

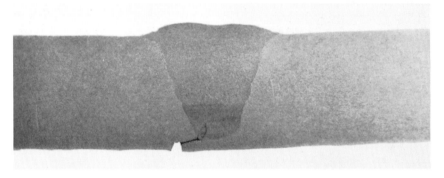

Fig. 3-29. Cracking starting from lack of fusion in root of weld of pipe butt joint with overlap root preparation.

lem for the welder to fuse properly into the zone formed between the two pipe edges. If proper fusion is not obtained, a notch will be retained. During subsequent service, this notch has led to cracking, as shown in Fig. 3-29.

Although the proper use of backing rings has not presented a major problem in power plant piping, smooth internal contours are often considered essential in chemical plants, refineries, nuclear power plants, beverage processing plants, and pharmaceutical plants. Thus, the inert-gas tungsten-arc welding process has been increasingly applied since 1950 to eliminate the need for backing rings.

Some of the fine cracks that may occur in the root pass made by the inert-gas tungsten-arc welding process are not always detected by radiographic, dye penetrant, or magnetic particle inspection techniques. Figure 3-30 illustrates the cross section of a weld containing

Design Considerations 39

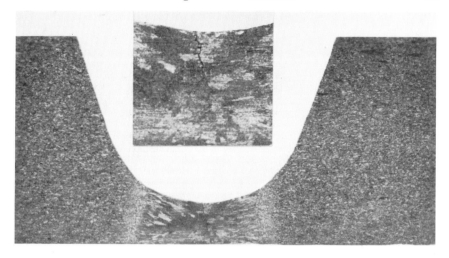

Fig. 3-30. Example of centerline crack in weld deposit not detected by radiographic inspection with x rays or radioactive isotopes. (Magnification 5×) (12× for inset)

Fig. 3-31. Crack in weld root pass in circumferential 1¼% Cr–½% Mo pipe butt weld made by inert-gas tungsten-arc welding. Crack does not penetrate through thickness of root weld deposit. (Magnification 5×)

a center-line crack that was not detected by radiographic inspection. Only deep etching of a cross section containing the tightly closed crack revealed its presence. Sometimes a crack does not even penetrate the full thickness of the root pass, as shown in Fig. 3-31. In such cases, the crack cannot be detected by dye penetrant or magnetic particle inspection. Lack of root penetration also may not be apparent in radiographic inspection, particularly when the wall thickness exceeds 1 in.

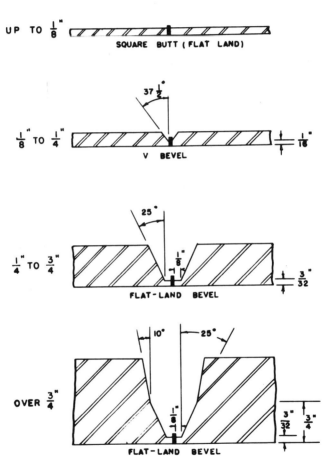

Fig. 3-32. Sketch of recommended weld end preparation providing greatest assurance for sound root-pass welds in critical piping.

Proper Weld Joint Design

Since even stringent inspection is not a guarantee that a weld joint is completely sound, the importance of selecting the joint preparation most advantageous for the welder and for the material cannot be overemphasized.

In piping and tubing, the joint designs that minimize the root pass cracking tendency when welding is done by the inert-gas tungsten-arc process employ flat-land and U-bevel preparations, as shown in Fig. 3-32. Although these preparations are somewhat more costly than standard 37½ degree V-bevel preparations, they are preferable in view of the shortcomings of inspection techniques. Consumable insert rings are also beneficial in insuring weld root soundness.[16, 17]

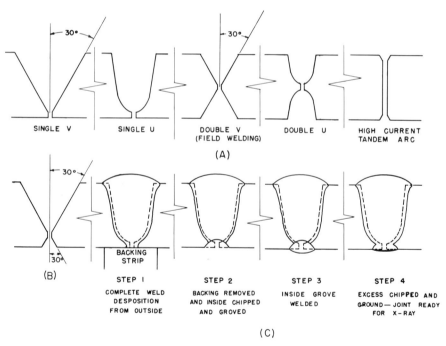

Fig. 3-33. Examples of conventional types of welding grooves used on pressure vessels. (A). Joint designs normally followed with submerged-arc welding. (B). Joint design for shielded metal-arc welding. (C). Steps generally followed in submerged-arc welding single U-grooves.[19]

42   Defects and Failures in Pressure Vessels and Piping

Other examples of good joint designs applicable with different welding processes to presure vessels and piping are shown in Figs. 3-33 and 3-34, respectively. These Figures illustrate only a few examples, as many other joint designs may be equally as suitable. The important point here is that the joint design selection should include consideration of the particular welding process used, the materials involved, and welding process variables, and should make it as easy as possible for the welder or welding equipment operator to produce a sound weld. For example, in gas-shielded consumable metal-arc welding of pipe butt welds, the $\frac{1}{8}$-in. thick land (Fig. 3-27) is more susceptible to contain insufficient fusion in the root (Fig. 13-8) than the feather edge root (Fig. 3-34).

Frequently, the susceptibility toward failure in a poorly designed weld joint is aggravated by excessive rigidity in a piping system. When it is known or suspected that the service involves mechanical or thermal fatigue, it becomes particularly important to design the system to avoid sharp notch conditions.

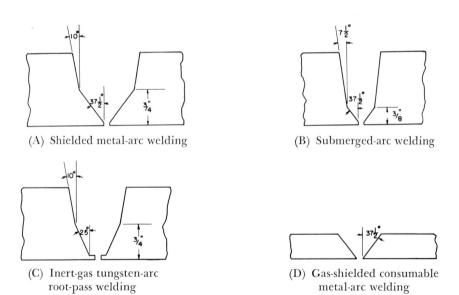

(A) Shielded metal-arc welding

(B) Submerged-arc welding

(C) Inert-gas tungsten-arc root-pass welding

(D) Gas-shielded consumable metal-arc welding

Fig. 3-34. Examples of pipe weld end preparations preferred for different welding processes.

Design Considerations

Whereas this is obvious in butt welds, in nozzle joints subject to mechanical fatigue, a flush weld may not exhibit as long a fatigue life as a joint in which the nozzle protrudes into the vessel. The effects of nozzle protrusion and weld penetration are illustrated in Fig. 3-35.[18] Applicable fatigue test data are given in Table 3-1.

TABLE 3-1. RESULTS OF PULSATING PRESSURE TESTS ON THE PRESSURE VESSEL BRANCH CONNECTIONS SHOWN IN FIG. 3-35.[18]

Test Limits: Max. pressure 2000 psi
              Min. pressure 100 psi
Mean shell hoop stress at max. pressure 18,200 psi

Design Detail	Branch	Weld Penetration	Life, Thousands of Cycles
1	Protruding	Partial	51.2 54.8 114.0
2	Flush	Partial	33.3 32.8 40.8
3	Protruding	Complete	136.8 180.7 203.1
4	Flush	Complete	23.4 56.0 75.7

Joint designs that make electrode manipulation and observation difficult are obviously more likely to result in joints that contain defects than those that do not.

With increasingly severe service conditions, it is essential that engineers designing vessels, tanks and piping systems and preparing fabrication specifications recognize the design conditions that tend to promote service failure. Too many designs and specifications that are inherently susceptible to cracking and failure are still being followed.

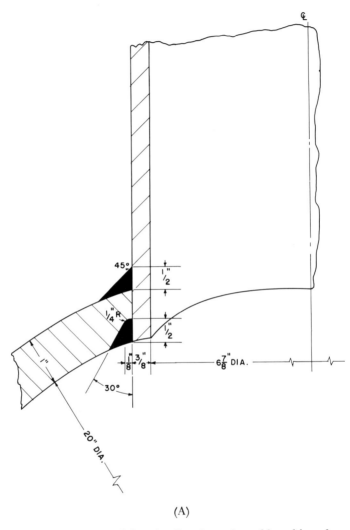

(A)

Fig. 3-35. Weld joint details of nozzle welds subjected to fatigue testing.[18]

Design Considerations

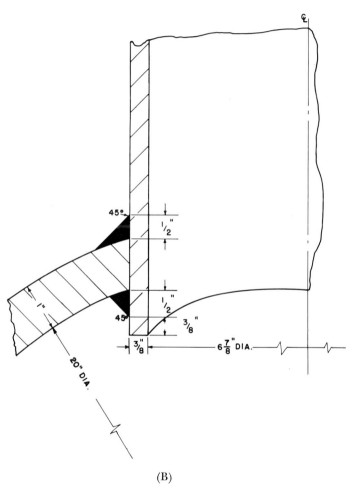

(B)

Fig. 3-35 cont'd

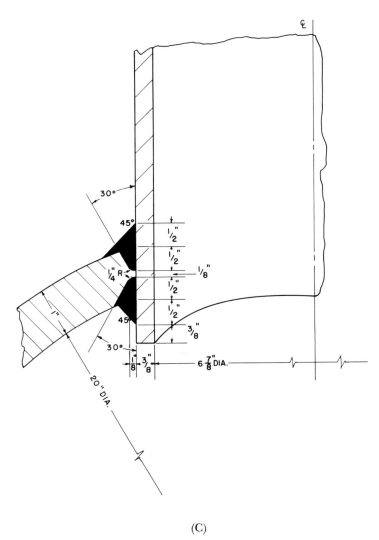

(C)

Fig. 3-35 cont'd

Design Considerations

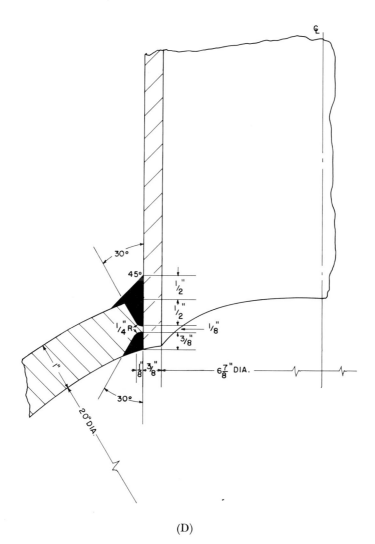

(D)

Fig. 3-35 cont'd

REFERENCES

1. Macfarlane, D. S., "Fatigue Strength of Gusseted Pipe Bends," *British Welding J.*, **9**, 659–669 (1962).
2. Anon., "Observations on Fractures of Welded Water Tanks of Locomotive Tenders," *Welding Research Abroad*, **7**, 77–81 (April, 1961).
3. Rankin, A. W., "Design and Fabrication of Steam Piping," *Welding J.*, **30**, 508–522 (1951).
4. International Institute of Welding Document XI-2-56 of Commission XI, "Pressure Vessels, Boilers and Pipe-Lines" (1956).
5. Wells, A. A., Lane, P. H. R., and Rose, R. T., "Stress Analysis of Nozzles in Cylindrical Pressure Vessels," Symposium on Pressure Vessel Research towards Better Design, The Institution of Mechanical Engineers, p. 17, London (1962).
6. Lane, P. H. R. and Rose, R. T., "Comparative Performance of Pressure Vessel Nozzles Under Pulsating Pressure," Symposium on Pressure Vessel Research towards Better Design, The Institution of Mechanical Engineers, p. 67, London (1962).
7. Soete, W., Hebrant, F., Dechaene, R., and Heirman, J., "The Reinforcement of Openings in Pressure Vessels," *Rev. Soudure—Brussels*, **17**, 19–34 (Jan. 1961).
8. Kooistra, L. F. and Lemcoe, M. M., "Low Cycle Fatigue Research on Full-Size Pressure Vessels," *Welding J.*, **41**, Res. Suppl., 297s–306s (1962).
9. Welter, G. and Dubuc, J., "Fatigue Resistance of Simulated Nozzles in Model Pressure Vessels of T-1 Steel," *Welding J.*, **41**, Res. Suppl., 368s–374s (1962).
10. Horseman, R. W., "Stresses in Oblique Nozzles on Pressure Testing of Reactor Pressure Vessels," Symposium on Pressure Vessel Research towards Better Design, The Institution of Mechanical Engineers, p. 43, London (1962).
11. Clarke, J. S. and Northup, M. S., "Design of Vessels for Petroleum Refining and Petrochemical Services," *Trans. ASME*, **86**, Series A, 411–418 (1964).
12. Marfels, W., "Methodik bei der Untersuchung von Schäden an geschweisten Bauteiler," *Schweissen und Schneiden*, **8**, 269–277 (1956).
13. Verein. d. Grosskesselbesitzer, 1960/61 Review.
14. Haack, "Werkstoffuntersuchungen an einem zerknallten Flüssiggas-Behälterwagen," *Schweissen und Schneiden*, **11**, 360–364 (1959).
15. Private Communication, Bayerwerk, A. G., Germany.
16. Thielsch, H., "Engineering Aspects of Inert-Gas Tungsten-Arc Welding of Piping," *Welding J.*, **34**, 1185–1195 (1955).
17. Thielsch, H., "Consumable Insert Rings for Better Pipe Welds," *Metal Prog.*, **84**, 91–93 (Sept. 1963).
18. Lane, P. H. R., "Pulsating Pressure Fatigue Tests on Pressure Vessel Branch Connections," *British Welding J.*, **5**, 327–332 (1958).
19. "Pressure Vessels and Boilers," p. 85.14, "Welding Handbook," 4th Ed., Section V, American Welding Society, New York, 1962.

chapter 4

MATERIAL SELECTION

GENERAL CONSIDERATIONS

Thirty years ago, when service requirements were generally less critical than those in industry today, carbon steel materials were normally adequate for the vast majority of applications involving pressure vessels, tanks, and piping in steam power plants, chemical plants, refineries, and other industrial plants. This has changed radically. A large variety of ferrous and nonferrous materials containing various alloying elements are now used in the manufacture of plate, piping, and tubing materials. Thus, the selection of materials is no simple task.

Numerous ferrous and nonferrous materials are recognized in the specifications of engineering and trade associations. The specifications most widely used are those prepared by the American Society for Testing and Materials, the American Petroleum Institute, and the American Society of Mechanical Engineers.

While the prevention of failures may be dependent on the material selected, one must not lose sight of economic considerations. In the selection of the proper materials, the most economical material that meets the requirements of the service conditions and of the applicable codes and standards should be specified. Many engineers who specify materials do not give sufficient consideration to temperature, pressure, stress levels, restraints resulting from supports or hangers, corrosion, section size and shape, thermal fatigue or shock, mechanical fatigue, etc. To play it safe, they often select a material far more costly than is necessary.

For example, stabilized Type 347 stainless steel piping has been specified for many installations involving only relatively mild corrosive conditions. In these cases, Type 304 stainless steel piping would have been adequate. Some piping components for nuclear power plants are currently fabricated of high nickel alloy or the costlier stainless steel materials. In many cases, more economical grades of stainless steel, low-alloy steel, or even carbon steel would be equally satisfactory.

LABORATORY TEST RESULTS

One fallacy that has resulted in critical service failures is the selection of materials for vessels and piping components on the basis of results obtained from small-scale laboratory test specimens. Tensile, creep, and impact tests are often conducted in this way. Small scale tests can be misleading in that they do not consider chemical and structural variations, and reflect the effects of shop fabrication, field welding, and service environments on large structures.

The precise control of the heat treatments given small test specimens may not be possible on pressure vessels, tanks and larger diameter sections of pipe, even when automatically controlled heat treating furnaces are used.

The selection of a base metal is most difficult when service conditions are somewhat more severe than those for which experience data are available. Frequently, the selection is then based on the interpretation of results obtained in accelerated mechanical or laboratory corrosion tests.

Considerable effort may be expended to simulate service conditions as closely as possible. Nevertheless, the results of the simulated tests frequently differ considerably from subsequent service experience. The difficulty, of course, is that the specimens used in the simulated tests usually are of simpler design, are far better machined to eliminate surface defects, and are much smaller than the fabricated assembly that will be exposed to the critical service conditions.

Material Selection

Generally, a test is accelerated by increasing the temperature and/or loads in order to reduce the period of time necessary to obtain the laboratory results. In creep tests, for example, periods of 1000, 5000 and 10,000 hours are normally employed. During subsequent service, of course, failures may develop after longer periods involving 50,000, 100,000 or 200,000 hours.

CARBON STEEL MATERIALS

Carbon steel plate and pipe materials are covered by various specifications, which recognize different levels of quality. For example, for the more critical applications, carbon steel pipe is often specified under ASTM specification A106, which recognizes only seamless piping, deoxidized with silicon. Its metallurgical microstructure is normally uniform, as illustrated in Fig. 4-1.

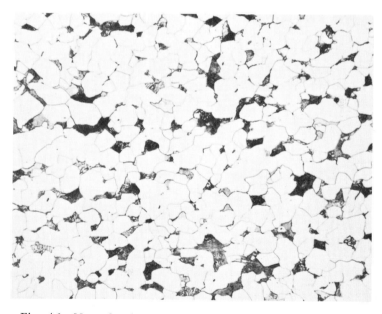

Fig. 4-1. Normal microstructure of seamless carbon steel pipe manufactured to the requirements of ASTM Specification A106, Grade B. (Magnification 100×)

52   *Defects and Failures in Pressure Vessels and Piping*

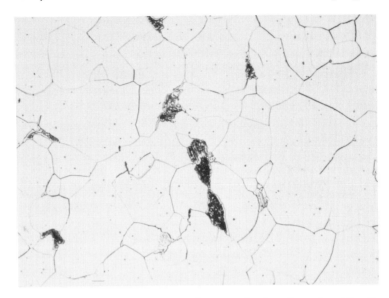

Fig. 4-2. Microstructure of carbon steel pipe (not deoxidized) manufactured to the requirements of ASTM Specification A53, Grade B illustrating large grain size and low carbon content. (Magnification 100×)

Fig. 4-3. Cross section of welded carbon steel pipe manufactured to the requirements of ASTM Specification A53 with smooth outside surface and weld notch along inside surface. (Magnification 7×)

Material Selection

Carbon steel pipe made to the requirements of ASTM Specification A53 is sold at a lower price. If A53 pipe is not deoxidized, the microstructure will be less uniform than that of A106 pipe, as shown in Fig. 4-2. This pipe may be either seamless or welded. Sometimes the welds may contain notches along the inside of the pipe, as shown in Fig. 4-3. These notches are not apparent on the outside surface which, in fact, may appear as smooth and uniform as seamless pipe. Nevertheless, A53 piping is adequate in many applications. In some cases it may even be more than adequate, and A120 pipe may be completely satisfactory.

However, there have been some failures in sufficiently critical service in A120 and A53 pipe which could have been avoided with A106 pipe. This includes applications where A53 and A106 pipes were used in the same piping systems and failures occurred only in the A53 material. Similar examples can be cited for corresponding plate materials. The majority of these failures have been associated with notch brittleness. Some have involved mildly corrosive environments or thermal or mechanical fatigue.

GRAPHITE FORMATION, EMBRITTLEMENT, AND CRACKING

Service Experience

Well known examples of failures that were not expected on the basis of laboratory studies are those stemming from graphitization of carbon and carbon-molybdenum steel piping after service at temperatures above 800°F.

In January 1943, the initial brittle failure occurred in a steam power plant.[1] Several subsequent failures have occurred in steam power plants [2,3] and refineries.[4]

Figure 4-4 illustrates a crack in a pipe spool section in the heat-affected zone parallel to the weld of a ½Mo steel pipe. The crack originated on the outside and has penetrated about 90% of the wall thickness. Severe graphitization which precedes cracking is shown in Fig. 4-5. In the final stages of graphitization, the steel in the affected zone will consist essentially of pure iron and carbon. Both phases are extremely brittle. Such degree of graphitization will severely reduce the ductility in the heat affected zone, as it is apparent from

54 *Defects and Failures in Pressure Vessels and Piping*

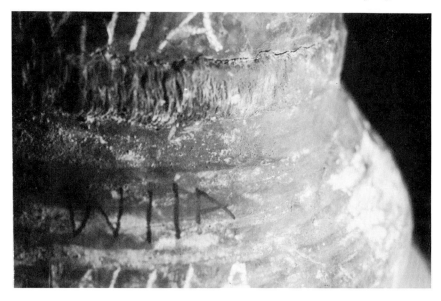

Fig. 4-4. A crack in a pipe spool section in the heat-affected zone parallel to the weld of a ½ Mo steel pipe.

the bend specimens shown in Fig. 4-6. Figure 4-7 illustrates cracking through a severely graphitized zone.

Graphite formation and the resulting embrittlement are generally limited to a narrow area in the heat-affected zone adjacent to the weld, normally about $\frac{1}{16}$ in. from the weld. Figure 4-8 illustrates the cracking in a valve parallel to the weld. Some forgings, however, can also exhibit several zones of severe graphitization and embrittlement, as shown in Fig. 4-9.

Graphite in steel is free carbon, which under some conditions forms by the disassociation of cementite into iron and carbon. The tendency for cementite to become unstable occurs primarily when steel has been heated very briefly at temperatures above the Ac_1 transformation temperature, where the steel changes from ferrite to austenite, and the iron-carbides (cementite) go into solution. This brief heating above the Ac_1 temperature, in effect, is produced by welding. It occurs in the base metal at a distance of about $\frac{1}{16}$ to $\frac{3}{32}$ in. from the weld. This, then, becomes the zone most susceptible to graphite formation.

Material Selection 55

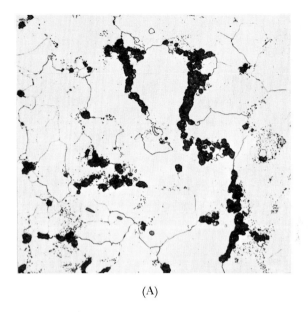

(A)

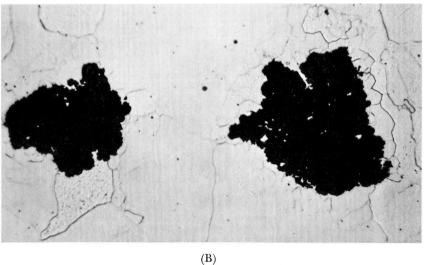

(B)

Fig. 4-5. Photomicrographs illustrating graphitization in pearlite grains still containing spheroidized carbides (A). Upon completion of the graphitization process the steel consists essentially of graphite and iron (B). (Magnifications 375×)

56   *Defects and Failures in Pressure Vessels and Piping*

Fig. 4-6. Typical cracking in bend test specimens in the weld-heat-affected zone upon bending by 5 to 15° indicating severe embrittlement due to graphite formation of $\frac{1}{2}$ Mo base metal.

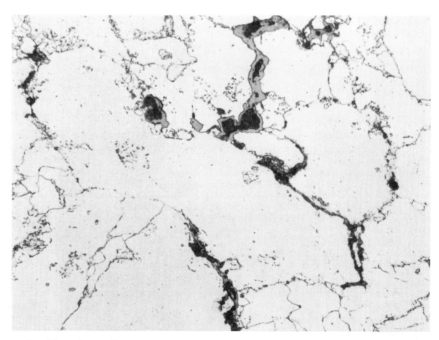

Fig. 4-7. Photomicrograph illustrating severe degree of graphitization and oxide filled cracks in $\frac{1}{2}$ Mo pipe base metal. (Magnification 500×)

Material Selection

Fig. 4-8. Typical cracking in heat-affected zone of ½ Mo valve base metal after service at temperatures over 850°F.

Fig. 4-9. Bend test specimen from forged flange of ½ Mo steel after slight bending illustrating four parallel cracks in zones containing severe graphitization.

Effects of Steel-making Practice

Steel-making practice has a significant bearing on the susceptibility of carbon and ½Mo steel to graphite formation. The addition of aluminum for deoxidation in quantities greater than 1 lb aluminum per ton of steel was generally involved in these steels. This aluminum deoxidation was common in the United States in the 1940's. One valve manufacturer utilizing titanium for deoxidation has not experienced graphite formation in his products. Similarly, molybdenum steels deoxidized with silicon, as has become common practice, have generally not been found to be susceptible to graphite formation.

Steel with ½Mo produced in the 1940's in England, where less than ½ lb of aluminum per ton of steel was used, have also not suffered from significant graphite formation.

Molybdenum steels with over 1Mo, which have also been occa-

58   *Defects and Failures in Pressure Vessels and Piping*

sionally produced, have not been reported to have formed graphite in service at temperatures over 800°F, even when deoxidized with 1 to 2 lb of aluminum per ton of steel. The Cr-Mo low-alloy steels, now extensively used at temperatures over 850°F, have not been found to form graphite.

Metallographic Grading of Graphitization

Metallographic evaluation is most widely used, to determine the degree of graphitization and to confirm the results of the guided-bend tests evaluating loss of ductility. The actual practice is to examine either the whole sectioned weld-probe specimen under the metallographic microscope or to cut smaller segments from this specimen representing heat-affected zone, base metal, and weld metal. An initial examination is made at 100 diameters to determine the presence and amount of graphitization. The specimen is subsequently examined at 500 diameters to establish more clearly the degree and type of graphitization. Both magnifications are necessary for the proper interpretation of graphitization.

Figure 4-10 illustrates a convenient metallographic grading system.[5]

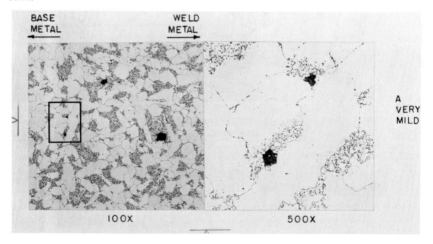

Fig. 4-10. Microstructures illustrating various degrees of graphitization: (A) "very mild" graphitization; (B) "mild" graphitization; (C) "moderate" graphitization; (D) "heavy" graphitization; (E) "severe" graphitization; (F) "severe" graphitization; (G) "severe" graphitization. (The original magnifications of 100× and 500× have been reduced to 60× and 300×, respectively).

Material Selection

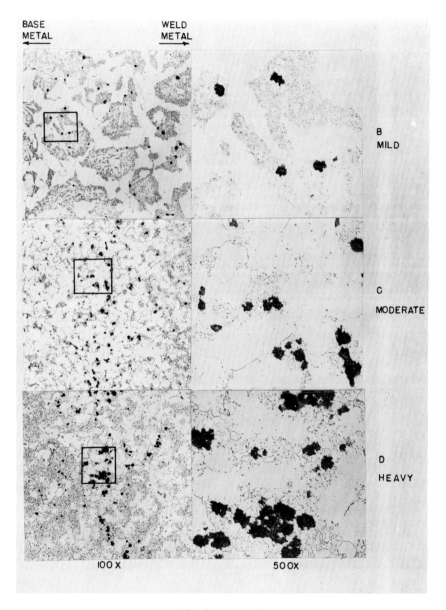

Fig. 4-10 cont'd

60 *Defects and Failures in Pressure Vessels and Piping*

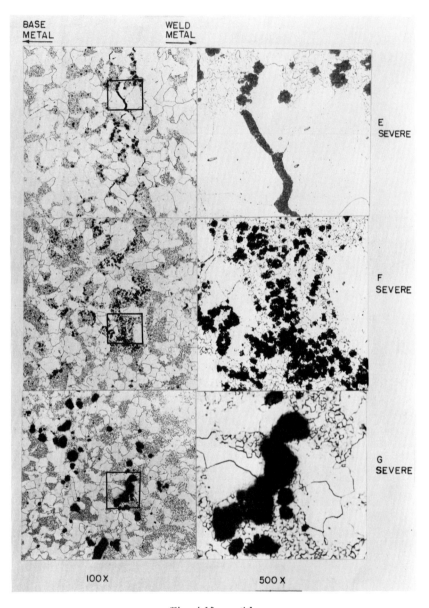

Fig. 4-10 *cont'd*

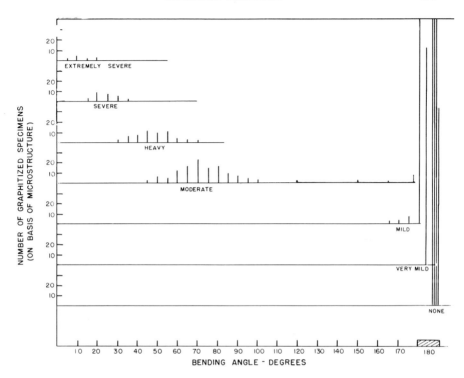

Fig. 4-11. Comparison of correlation between metallographic interpretation and bend test results of some 1700 weld-probe specimens.

The severity of graphitization depends upon the distribution, size and shape of the graphite particles. Whereas chain-type graphite, shown in Fig. 4-10E, usually indicates a "severe" or "extremely severe" condition, concentration of small nodular graphite particles (Fig. 4-10F) or a smaller number of very large nodular particles (Fig. 4-10G) may be just as severe as the chain-type formation.

The agreement between a careful metallographic evaluation and guided-bend tests is usually quite good. This is illustrated in Fig. 4-11 in which are summarized statistically the results of various commercial investigations. By far, the largest number of specimens exhibit either no graphitization or only very mild, mild, or moderate degrees of graphitization. The microstructures shown in Fig. 4-10 normally correspond to the bend angles listed in Table 4-1.

TABLE 4-1. CORRELATION BETWEEN METALLOGRAPHIC INVESTIGATIONS AND BEND TEST RESULTS.

Metallographic Evaluation	Bend Angle, Deg.
None	180
Very mild	180
Mild	90 to 180
Moderate	50 to 90
Heavy	30 to 50
Severe	15 to 30
Extremely severe	Below 15

Some variation may occur in the degree of graphitization around the pipe circumference or between the outside and inside surfaces of the pipe. The differences tend to be greater in cast and forged materials than in wrought pipe, as is illustrated in Table 4-2 and in Fig. 4-12. Only joints containing graphite were considered in Table 4-2. Degrees included are very mild, mild, moderate, heavy, severe and extremely severe.

TABLE 4-2. VARIATIONS IN DEGREE OF GRAPHITIZATION IN CARBON-MOLYBDENUM JOINTS SAMPLED AT TWO OR THREE LOCATIONS AS DETERMINED BY METALLOGRAPHIC TESTS.

Structural Component	Separation in Grading					
	Same	1 deg	2 deg	3 deg	4 deg	5 deg
Wrought pipe, %	50	30	17	3	0	0
Cast valve, %	30	25	20	15	7	3

Rehabilitation of Graphitized Weld Areas

Detection of the presence of graphite in its early stages in a weld joint may permit a rehabilitation program [6] consisting of a special heat treatment only. When graphitization is detected in a more critical degree, a much more costly rehabilitation program may be required which may either consist of gouging out the respective weld and heat-affected zones and rewelding, or it may even require the complete replacement of the pipe, valve, or header materials in the piping system.

Material Selection 63

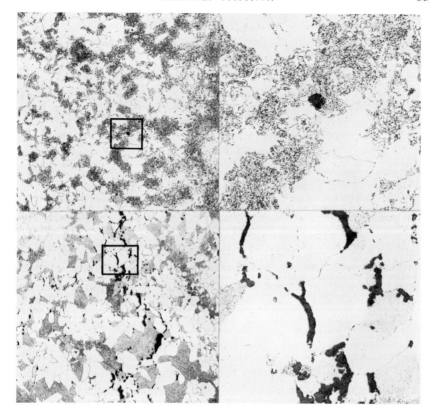

Fig. 4-12. Example of extreme variation in the degree of graphitization in two specimens removed 180° apart from the same valve joint involving ½ Mo steel. (Magnifications 70 and 350×)

In carbon-molybdenum steel materials, joints with degrees of graphitization representing "mild" or "moderate" can generally be rehabilitated by a solution heat treatment. Such a solution heat treatment normally consists of heating the weld and heat-affected zone areas for 2 hours at about 1750°F. This should be followed by slow cooling to room temperature and a subsequent "metallurgical stabilization" heat treatment involving 4 hours at 1300°F (±25°F).

Since it is generally believed that solution heat treatments are effective only where the degree of graphitization is "mild" or "moderate," solution heat treatment should not be applied to joints having formed "heavy" or "severe" degrees of graphitization. There is considerable evidence that where the graphite particles were extremely large, the 2-hour solution heat treatment at 1750°F is not sufficient to cause complete solution of the graphite and may, in fact, leave voids. Complete bend ductility may also not be restored, thus evidencing residual embrittlement. Particle size, therefore, enters into the interpretation of the type of rehabilitation program considered most advisable.

Current Practice

Wilson,[4] in analyzing refinery experience with plate materials, pointed out that ASTM A201 mild steel was considerably more susceptible to graphitization at temperatures over 800°F than ASTM A285 mild steel.

Because of the past propensity of certain types of carbon and carbon-molybdenum steels to lose ductility through graphitization at high temperatures, these are no longer extensively used in service at temperatures exceeding 750 and 800°F, respectively. Actually, the steels generally susceptible to graphitization were deoxidized with aluminum. The steels produced now are generally deoxidized with silicon. They are not considered susceptible to graphite formation after prolonged exposure to elevated temperatures. Stress relieving for 4 hours at temperatures about 150°F higher than the 1200°F generally used 10 to 15 years ago helps to suppress graphite formation in carbon-molybdenum steels, even when they are deoxidized with aluminum.

Although many laboratory studies have been conducted to evaluate the formation of graphite in carbon and carbon-molybdenum steels, these studies have not been able to reproduce the same critical type of graphite formation that has caused embrittlement and cracking in piping installations in high temperature service.

FERRITIC ALLOY STEELS IN HIGH TEMPERATURE SERVICE

Vanadium Steels

A number of instances of cracking have occurred in Mo and Cr-Mo steels alloyed with vanadium for increased creep strength at temperatures in the 800 to 1050°F range. Figure 4-13 illustrates cracking in the 1Mo–$\frac{1}{4}$V base metal adjacent to the weld.

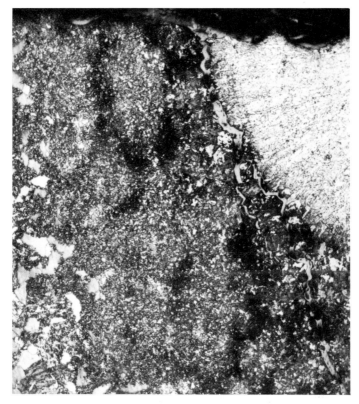

Fig. 4-13. Cracking in heat-affected zone of 1 Mo–$\tfrac{1}{4}$ V steel casting adjacent to weld. (Magnification 70×)

This cracking is normally associated with insufficient ductility, particularly in the heat-affected zone and along the weld. The vanadium carbide V_4C_3 which forms results in precipitation hardening. The effects on the creep strength and ductility are also influenced by the presence of other elements as carbon, molybdenum and chromium, and by heat treatment. Residual welding stresses and stresses caused by thermal fatigue may lead to cracking, particularly in materials of heavier wall thicknesses.

Vanadium-bearing steels (0.10–0.18C, 0.3–0.6Mn, 0.50–0.65Mo, 0.25–0.35V) used in steam piping in Europe [7,8] have been found to exhibit somewhat erratic notch toughness characteristics as well as susceptibility to weld metal fissuring.[8] As a result, one steam plant

66   *Defects and Failures in Pressure Vessels and Piping*

utilizing the steels is checking butt weld areas for cracking and embrittlement after operating periods of 1000 hours.[8]

In the welding of low Cr–Mo alloy steels containing vanadium, the susceptibility to cracking is less when vanadium-free filler metals are used such as 2¼Cr–1Mo.[9] This is asociated with a greater ductility because of the absence of vanadium in the weld deposit.

STAINLESS STEEL FOR CORROSION SERVICE

Type 304 is the most economical, austenitic stainless steel material available for pressure vessels, tanks and piping. In the seamless or welded condition, Type 304 provides excellent resistance to many corrosive environments. Some corrosive liquids, however, do attack this type significantly. In such cases, the selection of other stainless steel types or other ferrous or nonferrous materials is required.

Certain corrosive solutions attack Type 304 and other stainless steel types, such as Types 316, 317, 309 and 310, along the grain boundaries. This is usually associated with chromium-carbide precipitation when the steel is heated to between 900 and 1500°F. The intergranular chromium carbide precipitation may be removed by an annealing heat treatment at about 1950°F. This causes the chromium carbides to go into solution. It is referred to as *solution heat treatment*. During welding, an area of the base metal adjacent to the weld is heated into the 900 to 1500°F range. Those corrosive solutions that attack the intergranular chromium carbides are likely to attack the steel in the heat-affected zone, as illustrated in Fig. 4-14. To avoid this, either the extra low-carbon types of stainless steel

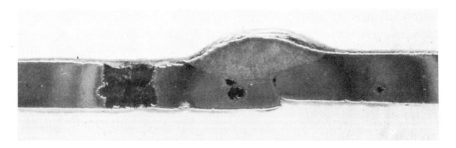

Fig. 4-14. Corrosion through heat-affected zone area in Type 304 stainless steel pipe caused by intergranular carbide precipitation.

Material Selection

should be specified, or grades should be used in which the carbides are stabilized with titanium (Type 321) or columbium (Type 347). Specifying that welding be followed by an immediate water quench does not eliminate this problem; it may actually be harmful to the weld quality obtained.

A number of examples of corrosion failures in stainless steel materials are illustrated in Chapter 17.

STAINLESS STEELS IN HIGH-TEMPERATURE SERVICE

Failures in Columbium-bearing Grades

In high-temperature service above approximately 1050°F, austenitic stainless steel materials are extensively used in pressure vessel and piping applications.

The danger involved in basing materials selection solely on the results of limited small-scale laboratory tests is illustrated by the substantial number of failures in Type 347 (18Cr–8Ni–Cb) stainless piping utilized in main steam piping in electric generating stations. Since the late 1940's, a number of steam power plants in the United States and Europe have been designed for steam temperatures of 1050 to 1100°F, and higher. Laboratory tests indicated that columbium stabilized 18Cr–8Ni stainless steel (Type 347) would be satisfactory. After 6 to 18 months service, however, the 6- to 12-in. diameter piping systems that were fabricated showed a significant tendency to crack in the heat-affected zones. This is illustrated in Fig. 4-15.

Since then, extensive laboratory investigations and test programs have been undertaken to determine the causes of the cracking.

The tendency toward cracking differed for different heats of steel. In some materials, cracking did not occur at all. Tests on specimens selected from pipe exhibiting good and poor service performance generally related to high and low hot ductility.[10,11,12] Variations in ductility in the base metal immediately adjacent to the heat-affected zone have been reported, also, when they were heated at the rates and temperatures occurring while welding.[13] The conditions of heating causing low hot ductility in the Type 347 stainless steel differed significantly between heats of steel within the normal limits of chemical composition produced commercially.[13,14]

68   *Defects and Failures in Pressure Vessels and Piping*

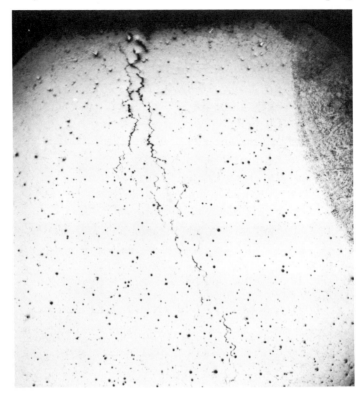

Fig. 4-15. Cracking in heat-affected zone of Type 347 stainless steel pipe joint in service at temperatures above 1050°F. (Magnification 35×)

The criterion for confirming the acceptability of specific heats of Type 347 is based on the recovery of ductility on cooling. Recovery after very little cooling is considered an indication of good resistance to cracking. Where considerable cooling is required, the material is considered susceptible to cracking.

Several causes have been suggested to explain the hot cracking. The most widely accepted explanation states that poor hot ductility is related to a low *liquation temperature*.[15-23] Liquation is explained as the melting primarily along grain boundaries well below the melting temperatures of the metal; i.e., it is hot shortness as described, also, elsewhere in this book. Columbium, carbon and nitrogen are major elements affecting the liquation temperature.[23] High carbon

Material Selection

and nitrogen in relation to columbium tend to raise the liquation temperature and increase resistance to cracking.[23]

It has not been clearly established if the heat-affected zone cracking in the base metal occurs during welding, or is the result of metallurgical changes caused in the heat-affected zone as a result of welding. The latter condition appears more likely, in that the brittleness is caused by a high degree of solution of columbium carbide at the high welding temperatures followed by reprecipitation in the presence of strain resulting from the thermal cooling stresses.[24-29] Figure 4-16 illustrates grain boundary precipitation of the columbium-rich phase and cracking in the heat-affected zone of Type 347 stainless steel pipe sections which failed in service.

Normal heats of Types 304, 316 and 321 stainless steels are generally not found susceptible to liquation and hot cracking. Nevertheless, some instances of liquation and hot cracking in 18Cr–10Ni–2½Mo and 18Cr–10Ni–0.3Ti base metal immediately adjacent to weld deposits have been reported.[56] High-temperature service failures of the type described in this section have not been observed in these materials. Thus, Type 316 stainless steel pipe has been used in several of the most recently constructed steam power plants operating at temperatures requiring stainless steel piping materials.

However, even in these materials, poor hot ductility can present a problem when ferrite is present, or is formed as a result of heating and straining during hot fabrication, forming, or testing.

Cracking in columbium-bearing stainless steels in steam plant service at temperatures of 1130 has also been reported in Germany, particularly in steels containing 0.07C–16Cr–21Ni–1½Mo–0.7V–0.7Cb, especially in heavier wall thickness.[30] The crack susceptibility varied in steels produced by different mills. On the other hand, there are some reports indicating that 16Cr–13Ni–Cb and 16Cr–13Ni–2Mo–Cb materials appear to have been satisfactory. Failures have not occurred in superheater tubing and steam piping at 1040 to 1200°F involving periods up to 10 years (1962) and sizes up to 4¼-in. OD by-0.80 in. wall thickness.[31] In flattening tests, the 16Cr–13Ni–2Mo–Cb material showed embrittlement after 75,000 hours at room temperature and 1110°F. However, the 16Cr–13Ni–Cb grade remained tough.[31] Possibly, the lower stress levels in the relatively light wall thicknesses involved have been primarily responsible for the apparent lack of

70 *Defects and Failures in Pressure Vessels and Piping*

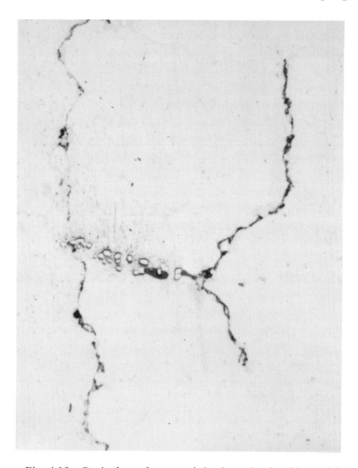

Fig. 4-16. Grain boundary precipitation of columbium rich phase and cracking in Type 347 stainless steel pipe section which failed in service. (Magnification 1000×)

major cracking in the 16Cr–13Ni–Cb materials, since laboratory tests nevertheless have shown that these materials are also susceptible to severe losses of hot ductility.[32, 33]

Sigma Phase in Stainless Steels

Sigma-phase formation may occur in austenitic stainless steels during long-time elevated temperature service between about 1050 and 1700°F.[34] The presence of considerable quantities of the sigma phase tends to result in a serious embrittlement at atmospheric temperatures. The temperature range and the rate of sigma formation may be affected to varying degrees by composition, metallurgical structure and deformation. For example, sigma forms much more slowly in the lower chromium fully austenitic 18Cr–8Ni alloys than in the higher chromium 25Cr–20Ni alloys. In fact, sigma will not appear in detrimental proportions in many 18Cr–8Ni grades.[35]

Although in the ferritic chromium stainless steels sigma may form in alloys containing as little as 17% chromium, considerable deformation usually is required if appreciable sigma-phase formation is to occur in reasonable service periods. However, ferritic stainless steels of higher chromium content, such as the 27% chromium alloy, may embrittle in as little as 100 hours at about 1100°F.

Although sigma formation occurs at temperatures above 1050°F, the embrittlement resulting from the presence of this phase becomes most serious after the steel has cooled to temperatures below about 500°F. Thus, at elevated temperatures, stainless steels in which the sigma phase has formed generally can withstand the normal mechanical or thermal fatigue and shock cycles which have been considered in the design. However, upon cooling to temperatures below about 500°F, the impact toughness is almost zero and maintenance of equipment containing embrittled steel when hammers or pipe wrenches must be used is impossible.[36] In order to remove the sigma phase and restore normal room-temperature ductility and toughness to the stainless steel, heat treatments above about 1900°F would be necessary. However, the effect of these treatments is only temporary and the sigma phase will reprecipitate when the steel is held in the temperature range in which sigma does form.

Since cold deformation of stainless steels suceptible to sigma formation very markedly increases the rates of sigma formation, the higher chromium stainless steels (25Cr–20Ni, 25Cr–12Ni) and ferritic chromium stainless steel (with over 17% chromium) should be placed into elevated temperature service exceeding 1050°F only after the cold formed sections have been annealed above 1700°F.

Actually, very few failures have been reported due to sigma formation. One example involved a Type 310 (25Cr–20Ni) stainless steel weld which failed at operating temperatures of 1500 to 1550°F with occasional swings to about 1700°F during the decoking operations.[37]

The hardness of the weld deposit was 228 to 246 Brinell and of the tube material 170 to 174 Brinell. A reverse bend snapped the section almost immediately, confirming very low ductility.

Another failure in a cyclone in an oil refinery made of 19Cr–12Ni–Cb stainless steel was found to be caused by sigma-phase formation. The resulting embrittlement led to cracking and was believed due to an excess of silicon and a deficiency of nickel (as low as 7.5%) in the stainless steel. Izod impact values in the embrittled area had decreased from 55–75 ft lb to 5 ft lb.[57]

Simulated Service Tests at Elevated Temperatures

As mentioned, data gathered from the testing of small-scale specimens may be misleading. However, the information gained from specimens of similar shapes and dimensions as the actual tube, pipe and vessel components in service and tested under conditions simulating those of actual service environments, can be quite helpful to designers. For new applications involving more critical temperatures and pressures than those on which long time service experience is available, simulated-service test data should be consulted in the selection of materials. For example, the data obtained from extensive tests now being conducted with various superheater tube alloys at steam temperatures of 1200, 1350 and 1500°F should be most helpful in the selection of boiler tube materials for temperatures of 1100°F and higher.[38, 39]

SELECTION OF WELDING FILLER METAL

Selection of the proper welding filler metal is extremely important in the prevention of pipe and vessel failures. In the 1930's and 1940's, welders generally were not much concerned about the particular steel electrodes that were used. One steel electrode was considered pretty much the same as any other. Because of the development of new base metal alloys, various types of electrode coatings, and different welding processes, and because of the increase in severity of service conditions, it has become extremely important that the proper filler metal composition is used.

Ideally, the weld deposit should be identical in mechanical prop-

Material Selection

erties and metallurgical structure to the base metal. Although a perfect match is never obtained, close similarities are possible with many commercial alloys. There are some alloys, however, that are not readily welded with some or even any of the welding processes. Welding filler metals of a somewhat different composition from that of the base metal may have to be selected in such cases.

For example, titanium stabilized Type 321 stainless steel piping is normally welded with columbium stabilized Type 347 stainless steel electrodes. If titanium were added to the electrode, most of it would be oxidized out in the welding arc, and there would be no stabilizing alloying element in the weld deposit. Columbium, on the other hand, is not lost in significant quantities during welding.

DISSIMILAR JOINTS

Slight Dissimilarities in Composition

Caution must be exercised in using welding electrodes of different compositions or with different metallurgical structure from that of the base metal, especially where service temperatures exceed 800°F. In most dissimilar metal combinations, changes take place across the fusion zone at these temperatures. (Chapters 8 and 13 contain examples of failures caused by very slight metallurgical differences between weld and base metals.)

Slight dissimilarities in composition can also be critical in corrosive service.

Figure 4-17 illustrates a corrosive failure in a Type 304 stainless steel joint welded with Type 308 stainless steel electrodes. The corrosive solution attacked the ferrite phase in the Type 308 weld deposit. The composition of the Type 304 base metal was higher than usual in nickel content and lower in chromium, thus providing an essentially fully austenitic metallurgical structure. If the weld metal had been Type 310, which normally is austenitic, the type of corrosive attack shown in Fig. 4-17 would not have occurred. Selection of Type 310 base metal, of course, would also have been desirable.

These considerations emphasize the importance of care in the selection of the welding filler metal. Moreover, welders and super-

74 *Defects and Failures in Pressure Vessels and Piping*

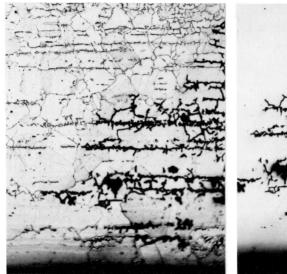
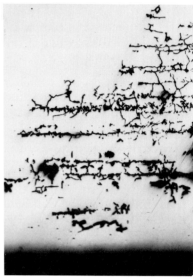

Fig. 4-17. Selective corrosion of ferrite phase in partially ferritic Type 308 stainless steel weld deposit in hot-water pipe line containing residual chlorides. (Magnification 120×)

visory personnel must be made cognizant of the hazardous consequences that can result from mixups of filler metals and improper handling.

The selection of all materials in a pressure vessel or piping system must be based on their properties and characteristics in the specific sizes and thicknesses involved, their behavior under *normal* fabrication procedures, and their ability to withstand service conditions for extended periods of time.

Potential Failures in Slightly Dissimilar Joints in Elevated Temperature Service

In the majority of cases of gradually developing failures in dissimilar metal joints, the cracking starts on the inside of the pipe wall. Detection, therefore, is extremely difficult, if not impossible.

Whereas bend tests are widely employed to evaluate the embrittlement in weld joints, they are not conclusive in determining the weakening across a dissimilar metal joint. The majority of bend bars removed from dissimilar metal joints, containing gradually propagating cracks near the root of the weld, have been bent satisfactorily to 180 degrees when removed just outside the cracked area. It is apparent, therefore, that the carbon migration leaves the joint area extremely notch-sensitive, but does not affect the bend ductility significantly.

A dissimilarity is likely to be more critical if different elements are involved than if different amounts of the same elements are involved. For example, at temperatures exceeding 800°F, the dissimilarity between unalloyed carbon steel and $\frac{1}{2}$Mo steel is more critical than the dissimilarity between $1\frac{1}{4}$Cr–$\frac{1}{2}$Mo and $2\frac{1}{4}$Cr–1Mo low-alloy steels.

In general, weld deposits are morse adversely affected by decarburization than wrought base metal.

Major Dissimilarities in Composition

A major dissimilarity consists of a weld joint having a chemical composition different by at least 5% of the total composition of one or several of the adjacent base metals.

Dissimilar-metal weldments are made when two or more different materials are joined by welding. For example, chromium-molybdenum alloy-steel piping may be welded to austenitic stainless-steel pumps, valves, or turbine casings.

Dissimilar-metal welds have also been used to join components of essentially the same composition. This has been done on some low-alloy steel piping materials where, by employing stainless steel filler metals (usually 25Cr–20Ni), postheat treatments have not been considered necessary.

A third important group of dissimilar-metal welds consists of weld overlays. For many service applications, pressure vessels and piping are being overlaid with highly alloyed or nonferrous materials to provide special corrosion-resisting, heat-resisting or wear-resisting properties.

Factors Causing Failure

In order for a failure to occur, stresses must be of sufficient magnitude to "tear" apart a material. The greater the "brittleness" of the material, the smaller are the stresses which will cause cracking. Brittleness usually is associated with low ductility and low toughness.

As creep resistance diminishes in the dissimilar-metal joint zone, the stress level at which cracking will occur also is reduced. This problem may become particularly critical at temperatures above 800°F where metallurgical changes across the dissimilar bond * may seriously reduce creep strength in the base metal, the weld metal adjacent to the bond, or in both.

In combinations between dissimilar steel compositions, such as low-alloy steels and stainless steels, the effect of carbon migration across the bond is of particular concern.[40-42] In general, carbon tends to migrate from the lower alloy steel composition into the more highly alloyed material. Carbon also tends to migrate in the direction from high total carbon content to a lower total carbon content.

The threshold stress level which causes cracking and crack propagation is a combination of various residual and load stresses. This combination includes stresses caused by differences in the coefficient of expansion between the dissimilar materials and, to a lesser extent, differences in the rate of heat transfer. In joints between austenitic stainless steels and ferritic mild or low-alloy steels, the greater coefficient of expansion of the austenitic stainless steels (approximately 30%) and the lower thermal conductivity (approximately 40%) may introduce stresses of considerable magnitude. Weld assemblies consisting of austenitic stainless steel weld deposits adjacent to low-alloy steels in service environments exposed to severe thermal fatigue or shock may be distorted. They also may fail due to the initiation and propagation of a crack in the ferritic alloy-steel material which tends to exhibit lower toughness and creep resistance. Figure 4-18 illustrates cracking around the periphery of a socket weld inadvertently made with Type 347 stainless steel electrodes and joining $1\frac{1}{4}$Cr– $\frac{1}{2}$Mo thermocouple wells to $1\frac{1}{2}$Cr–$\frac{1}{2}$Mo pipe. Only four months of service at 1000°F were involved in the initiation and propagation of the cracking.

In corrosive environments favorable to stress-corrosion cracking, the generally higher stresses across a dissimilar-metal joint tend to increase even further the susceptibility of the joint to stress corrosion and caustic embrittlement.

The presence of surface notches in the dissimilar-metal area fur-

---

* *Bond* is defined as the junction of the weld metal and the base metal.

Material Selection

Fig. 4-18. Cracking around periphery of Type 347 stainless steel weld deposit made between $1\frac{1}{4}$ Cr–$\frac{1}{2}$ Mo pipe and $1\frac{1}{4}$ Cr–$\frac{1}{2}$ Mo thermocouple tap.

ther raises the stress level. The consequent higher maximum stress at the root of the notch may be sufficient to permit crack initiation, which requires a greater stress level than the subsequent continued crack propagation.

Elevated Temperature Service

Numerous failures in service at high temperatures, generally over 800°F, are reported in the literature where low-alloy steel piping materials have been welded with austenitic stainless steel welding filler metals.[40,43-46]

As a result of these failures, it is considered good engineering practice in high-temperature service applications to select welding filler metals corresponding as nearly as is practical to the compositions of the base metals.

There are, however, many applications where dissimilar materials must be combined. Particularly in steam power plants involving temperatures above 1000°F, combinations between austenitic stainless steel components and ferritic low-alloy steel components are commonly required in superheaters, reheaters, main steam, and hot reheat piping. In such applications, the selection of the welding filler metal is particularly important. Extensive test programs involving

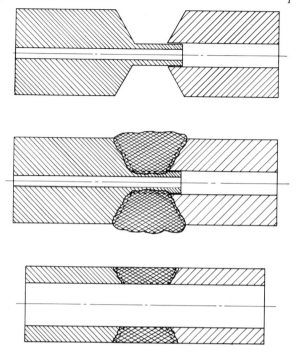

Fig. 4-19. Dissimilar-metal transition section used in several nuclear power plant applications, illustrating original sections and final appearance after welding, internal boring and outside machining to eliminate mechanical notch conditions.

such metal combinations, which readily produced failures where Type 347 and 310 filler metals were used,[44,48] have led to the application of nickel-base filler metals.[30,48-53] Nevertheless, some creep tests at 1025°F with specimens containing 75Ni–15Cr–1Mo type welds on ferritic alloy steels showed a significant loss of creep strength after 1000 hours.[30,51]

A material now commonly used is "Inconel" containing approximately 72Ni, 20Cr, 3Mo which has a coefficient of expansion between those of the low-alloy steels and stainless steels.[48] Experience appears to confirm these test results since to date major service failures have not been reported.

Of course, there exists also a size effect. In tubing of small diameters up to about 3-in. OD, dissimilar joints between austenitic stainless steel and low-alloy steel tubing have generally been performing satisfactorily, even when welded with austenitic stainless steel welding filler metals.

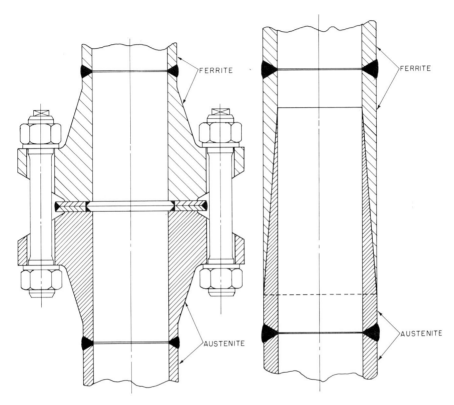

Fig. 4-20. Dissimilar-metal transition sections utilized in Germany in steam plant piping systems.[31, 49]

PREFABRICATED DISSIMILAR TRANSITION SECTIONS

For service involving high temperatures and thermal fatigue conditions, increasing use is being made of special prefabricated transition sections. By machining exterior and interior surfaces, mechanical notch conditions can be avoided.[47]

A prefabricated tubular transition section involving a combination between stainless steel and mild steel, "Monel" and stainless steel

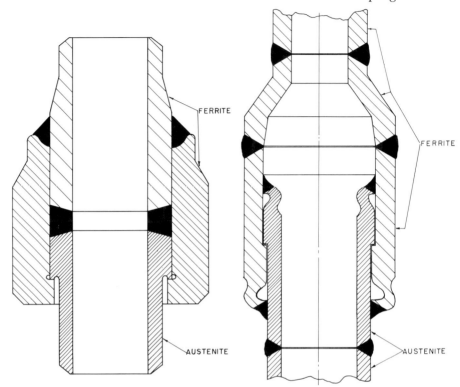

Fig. 4-20 cont'd

and others as used in a number of nuclear power plants is illustrated in Fig. 4-19.

Four different austenitic stainless steel to low-alloy steel transition sections successfully used in steam power plants in Germany are illustrated in Fig. 4-20.[31, 49, 54, 55]

In on site, erection, where welding conditions are not as perfect, welding of these transition sections then involves only similar materials.

REFERENCES

1. Emerson, R. W., "Carbide Instability of Carbon-Molybdenum Steel Piping," ASTM-ASME Symposium on Graphitization (1943); *Trans. ASME,* **66,** 5–15 (1944).
2. Nuchols, J. B. and McGuffey, J. R., "Graphitization Failures in Piping," *Mech. Eng.,* **81,** 42–45 (1959).

3. Mahoney, C. H. and Dritt, W. S., "Oak Ridge Reports on Graphitization in H-P Steam Lines," *Power Engineering*, **63**, 68–69 (Mar. 1959).
4. Wilson, J. G., "Graphitization of Steel in Petroleum Refining Equipment," Welding Research Council Bulletin No. 32 (Jan. 1957).
5. Thielsch, H., Phillips, E. M., and Jerome, E. R., Jr., "Considerations in the Evaluation of Graphitization in Piping Systems," *Welding J.*, **34**, Res. Suppl., 286s–294s (1955).
6. Thielsch, H., "Weld Probe Sampling Programs of Graphitized Piping Are Still Your Best Guarantee of Safety," *Heating, Piping & Air Conditioning*, **31**, 116–120 (Mar. 1959).
7. Wehrberger, F., "Erfahrungen beim Bau der Heissdampfleitungen aus dem Stahl 14 MoV 63 im Kraftwerk Nord des Volkswagenwerkes Wolfsburg," *Mitt. Ver. Grosskesselbesitzer*, **78**, 199–207 (1962).
8. Haferkamp, M., "Erfahrungen mit dem Werkstoff 14 MoV 63 bei Heissdampfrohrleitungen im Grosskraftwerk Mannheim," *Mitt. Ver. Grosskesselbesitzer*, **84**, 186–194 (1963).
9. Harris, P. and Lee, E. H., "Some Aspects of Welding in Steam Turbines," *British Welding J.*, **9**, 60–69 (1962).
10. Soldan, H. M. and Mayne, C. R., "Ductility Related to Service Performance of Heavy Wall Austenitic Pipe Welds," *Welding J.*, **36**, Res. Suppl., 141s–147s (1957).
11. Soldan, H. M. and Schnabel, G. J., Discussion of ASME Paper No. 62-Met.-11, "A Critical Analysis of the Weld Heat Affected Zone Hot Ductility Test," *Welding J.*, **42**, Res. Suppl., 55s–57s (1963).
12. Soldan, H. M. and Schnabel, G. J., "The Hot Bend Test—A Tool to Predict Suitable Weldability and High Temperature Service Performance of Austenitic Steels," *Welding J.*, **43**, Res. Suppl., 353s–357s (1964).
13. Nippes, E. F., Savage, W. F., and Grotke, G., "Further Studies of the Hot Ductility of High Temperature Alloys," Welding Research Council Bulletin No. 33 (Feb. 1957).
14. Nippes, E. F., Savage, W. F., Bastien, B. J., Mason, H. F., and Curran, R. M., "An Investigation of the Hot Ductility of High Temperature Alloys," *Welding J.*, **34**, Res. Suppl., 183s–196s (1955).
15. Rollason, E. C. and Bystram, M. C. T., "Hot Cracking of Austenitic Welds With Special Reference to 18/13/1 Cr-Ni-Cb Alloy," *J. Iron Steel Inst.*, **169**, 347–352 (1951).
16. Parks, J. M., "Discussion of Welding Type 347 Stainless Steel for 1100°F Turbine Operation," *Welding J.*, **34**, 568–570 (1955).
17. Puzak, P. P., Apblett, W. R., and Pellini, W. S., "Hot Cracking of Stainless Steel Weldments," *Welding J.*, **35**, Res. Suppl., 9s–17s (1956).
18. Puzak, P. P. and Rischall, H., "Further Studies on Stainless Steel Hot Cracking," *Welding J.*, **36**, Res. Suppl., 57s–61s (1957).
19. Emerson, R. W. and Jackson, R. W., "The Plastic Ductility of Austenitic Piping Containing Welded Joints at 1200°F," *Welding J.*, **36**, Res. Suppl., 89s–104s (1957).
20. Nippes, E. F., "The Weld Heat-Affected Zone," *Welding J.*, **38**, Res. Suppl., 1s–18s (1959).
21. Borland, J. C., "Generalized Theory of Super-Solidus Cracking in Welds and Castings," *British Welding J.*, **7**, 508–512 (1960).

22. Younger, R. N., Borland, J. C., and Baker, R. G., "Heat-Affected Zone Cracking of Two Austenitic Steels," *British Welding J.*, **8**, 575–578 (1961).
23. Cullen, T. M. and Freeman, J. W., "Metallurgical Factors Influencing Hot Ductility of Austenitic Steel Piping at Weld Heat-Affected Zone Temperatures," *Trans. ASME*, **85**, Series A, 151–164 (1963).
24. Truman, R. J. and Kirby, H. W., "Some Ductility Aspects of 18-12-1Nb Steel," *J. Iron Steel Inst.*, **196**, 180–188 (1960).
25. Irvine, K. J., Murray, J. D., and Pickering, F. B., "The Effect of Heat Treatment and Microstructure on the High Temperature Ductility of 18% Cr-12% Ni-1% Nb Steels," *J. Iron Steel Inst.*, **196**, 166–179 (1960).
26. Younger, R. N. and Baker, R. G., "Heat-Affected Zone Cracking in Welded High Temperature Austenitic Steels," *J. Iron Steel Inst.*, **196**, 188–194 (1960).
27. Asbury, F. E., Mitchell, B., and Toft, L. H., "Cracking in Welded Joints of Austenitic Steel in C. E. G. B. Power Stations," *British Welding J.*, **7**, 667–678 (1960).
28. Moore, N. E. and Griffiths, J. A., "Microstructural Causes of Heat-Affected Zone Cracking in Heavy Section 18-12-Nb Austenitic Stainless Steel Welded Joints," *J. Iron Steel Inst.*, **197**, 29–39 (1961).
29. Younger, R. N. and Baker, R. G., "Heat-Affected Zone Cracking in Welded Austenitic Steels During Heat Treatments," *British Welding J.*, **8**, 579–587 (1961).
30. Class, I., "Schlussfolgerungen für den Dampfkesselbetrieb—Neuere Beobachtungen und Erkenntnisse über das Verhalten von ferritischen und austenitischen Kesselstählen," *Mitt. Ver. Grosskesselbesitzer*, **80**, 326–337 (1962).
31. Ruttmann, W. and Brunzel, N., "10 Jahre austenitische Stähle im Kesselbetrieb," *Mitt. Ver. Grosskesselbesitzer*, **80**, 310–324 (1962).
32. Weinmann, W., "Über das Schweissen hochwarmfester Werkstoffe," *Industrie Anzeiger*, **82**, 1595–1601 (Nov. 18, 1960).
33. Kreitz, K., "Stähle für Energieanlagen," *Brennstoff-Wärme-Kraft*, **14**, 163–166 (1962).
34. Thielsch, H., "Physical Metallurgy of Austenitic Stainless Steels," *Welding J.*, **29**, Res. Suppl., 577s–621s (1950).
35. Nicholson, M. E., Samans, C. H., and Shortsleeve, F. J., "Composition Limits of Sigma Formation in Nickel-Chromium Steels at 1200°F (650°C)," *Trans. Am. Soc. Metals*, **44**, 601–619 (1952).
36. Rutherford, J. J. B., "Some Experiences in Service," High-Temperature Properties of Metals, American Society for Metals, Cleveland, Ohio, 133–170 (1951).
37. Resume of High-Temperature Investigations Conducted During 1951–1952, The Timken Roller Bearing Co., Canton, Ohio.
38. Clark, C. L., Rutherford, J. J. B., Wilder, A. B., and Cordovi, M. A., "Metallurgical Evaluation of Superheater Tube Alloys After 6 Months' Exposure at Temperatures of 1100 to 1500°F," *Trans. ASME*, **82**, Series A, 35–67 (1960).
39. Clark, C. L., Rutherford, J. J. B., Wilder, A. B., and Cordovi, M. A., "Metallurgical Evaluation of Superheater Tube Alloys After 12 and 18 Months' Exposure to Steam at 1200, 1350 and 1500°F," *Trans. ASME*, **84**, Series A, 258–288 (1962).

40. Thielsch, H., "Stainless-Steel Weld Deposits on Mild and Alloy Steels," *Welding J.*, **31**, Res. Suppl., 37s–64s (1952).
41. Cristoffel, R. J. and Curran, R. M., "Carbon Migration in Welded Joints at Elevated Temperatures," *Welding J.*, **35**, Res. Suppl., 457s–468s (1956).
42. Eckel, J. F., "Diffusion Across Dissimilar Metal Joints," *Welding J.*, **43**, Res. Suppl., 170s–178s (1964).
43. Weisberg, H., "Cyclic Heating Test of Main Steam Piping Joints Between Ferritic and Austenitic Steels—Sewaren Generating Station," *Trans. ASME*, **71**, 643–664 (1949).
44. Blaser, R. V., Eberle, F., and Tucker, J. T., "Welds Between Dissimilar Alloys in Full-Size Steam Piping," *Proc. ASTM*, **50**, 789–807 (1950).
45. Lien, G. E., Eberle, F., and Wylie, R. D., "Results of Service Test Program on Transition Welds Between Austenitic and Ferritic Steels at the Philip Sporn and Twin Branch Plants," *Trans. ASME*, **76**, 1076–1083 (1954).
46. Wiester, H. J., "Rissbildung an austenitisch geschweissten Kesseltrommeln," *Mitt. Ver. Grosskesselbesitzer*, **7**, 1 (1949).
47. Blumberg, H. S., and Bunn, W. B., Discussion of ASME Paper "Cyclic Heating Test of Main Steam Piping Joints Between Ferritic and Austenitic Steels —Sewaren Generating Station," *Trans. ASME*, **71**, 651–653 (1949).
48. Tucker, J. T., Jr. and Eberle, F., "Development of a Ferritic-Austenitic Weld Joint for Steam Plant Application," *Welding J.*, **35**, Res. Suppl., 529s–540s (1956).
49. Jahn, E., "Schweissverbindungen zwischen ferritischen und austenitischen Stählen in Versuch und Praxis," *Mitt. Ver. Grosskesselbesitzer*, **77**, 86–93 (1962).
50. Class, I., "Hochtemperaturbeanspruchte Rohrleitungen," Die Schweisstechnik in der Energieversorgung, *Fachbuchreihe Schweisstech.*, **17**, 80.
51. Brown-Boveri Company, Special Report, *Schweissung*, 1961/1962.
52. Kauhausen, E., Kaesmacher, P., and Sadowski, S., "Probleme der Schweissung von warmfesten Stählen," *Werkstatt Betrieb*, **93**, 653–661 (1960).
53. Slaughter, G. M. and Housley, T. R., "The Welding of Ferritic Steels to Austenitic Stainless Steels," *Welding J.*, **43**, Res. Suppl., 454s–460s (1964).
54. Class, I., "Stand der Entwicklung nichtlösbarer Verbindungen von ferritischen und austenitischen Stählen," *Mitt. Ver. Grosskesselbesitzer*, **60**, 181–207 (1959).
55. Schinn, R. and Ruttmann, "Thermoschockversuche an Austenit-Ferrit-Verbindern," *Mitt. Ver. Grosskesselbesitzer*, **57**, 407 (1958).
56. Williams, N. T. and Myers, J., "A Note on Sub-Surface Cracking of Austenitic Steels During Welding," *British Welding J.*, **9**, 432–435 (1962).
57. Pull, D. J., "Mechanical Engineering in a Modern Oil Refinery," *Proc. Inst. Mech. Eng.*, **176**, 495–521 (1962).

chapter 5

IMPROPER HANDLING AND PROCESSING

Improper processing, handling and storage involving pressure vessel and pipe materials and welding filler metals can also cause defects and, in some instances, result in service failures.

HANDLING OF PLATES, PIPING AND EQUIPMENT

Particularly in notch-brittle steels, it is important that handling by cranes, chain falls, or other equipment does not leave severe gouges or dents in the steel.

There have been several instances where the handling of steel plates by cranes has resulted in fracture of the plates. The causes have been the bending moments produced by the crane handling and the inherent notch brittleness of the steel.

Tong marks can also produce severe notches. Since tong marks can also cause hardening, notch brittleness may be further increased, and bursting strength reduced. For example, on a 5½-in. 17-lb N-80 casing, severe dingeing caused failure at a bursting pressure of 8640 psi instead of the estimated bursting strength of 14,370 psi. Thus, the combination of dingeing and die notching reduced the bursting strength by 40%.[1] Other instances have been reported where the bursting pressures of oil well casings were reduced as much as 70% with notch depths of 15 to 17% of the wall thickness and ovality of 3 to 4%.[2]

Improper Handling and Processing   85

Intermittent gouges produced on large diameter gas transmission piping by hardened steel crawlers on paint priming and coating machines have also caused pipe ruptures.[3]

Severe localized hammering of a 3-in. diameter killed ½Mo steel pipe increased the transition temperature sufficiently to result in a brittle fracture when the pipe was accidentally dropped.[4]

TRANSPORTATION OF VESSELS AND PIPE

A number of failures have been caused or initiated as a result of improper transportation.

Fatigue cracks have been caused in several pipes for gas-transmission service which had been riding in rail cars on rivet heads.[5] The pipe surface became gouged and dented, resulting in high localized stresses and fatigue cracking. Actual failures generally have occurred during pressure testing of the gas-transmission piping.[6]

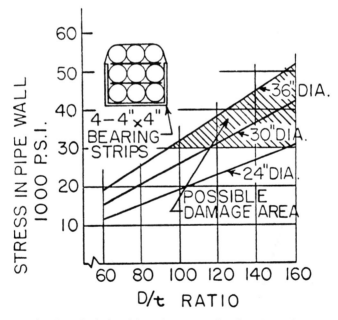

Fig. 5-1. Relationship of stresses developed during 2-g acceleration in railway cars for thin wall pipe.[6] (D = diameter, t = thickness).

Abrasion, gouging, fatigue, and end bevel damage of pipe have provided serious problems in rail transportation. In shipping thin wall pipe, too high a diameter (D) to thickness (T) ratio can result in excessive stresses and damage. This is illustrated in Fig. 5-1 for three pipe sizes. The calculated stresses at a 2 g-level of loading are the local bending stresses on the four 4-in. by 4-in. bearing strips across the bottom of railroad cars.[7]

PROCESSING OF WELDING ELECTRODES

Just because two or more lots of welding filler metals from the same or different manufacturers bear the same AWS-ASTM designation (for example, E8018-B2, which indicates $1\frac{1}{4}$Cr–$\frac{1}{2}$Mo electrodes with low hydrogen iron powder coatings and a minimum weld deposit tensile strength classification of 80,000 psi), does not mean that the weld deposits will be equally satisfactory. Improper processing by one manufacturer may have produced an electrode incapable of producing Code acceptable welds. Figure 5-2 illustrates porosity in a $1\frac{1}{4}$Cr–$\frac{1}{2}$Mo weld deposit from a commercially marketed electrode handled in accordance with quality conscious shop fabrication practices.

Electrode manufacturers from time to time change the composition of their welding filler metals and coatings. To avoid concern by their customers, changes in the coating within a specific classification often are not publicized. While such a change may not adversely affect the welding of plate, it may not be acceptable for welding pipe

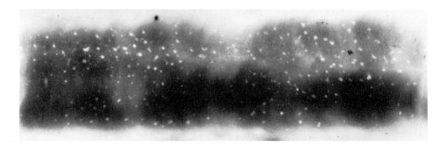

Fig. 5-2. Radiograph illustrating porosity in weld deposit made with a standard $1\frac{1}{4}$ Cr–$\frac{1}{2}$ Mo flux-covered electrode.

since rapid changes in electrode positions around the circumference of the pipe are involved.

STORAGE OF WELDING ELECTRODES

Many types of electrodes tend to absorb moisture in their coatings when exposed to humid atmospheres. Electrodes with low-hydrogen type coatings are particularly susceptible to moisture absorption. Such electrode types are extensively used for the welding of alloy steels, and, to a considerable extent, are used even on carbon steels.

Electrodes which have absorbed moisture have a tendency to produce porosity in weld deposits and, in some hardenable alloy steels, may cause fissuring and cracking.

Contamination of electrodes by moisture pickup is avoided by storage in stationary and portable electrode drying ovens.

REFERENCES

1. Texter, H. G., "Why Oil-Well Tubing and Casing Fail. Why They Burst, Leak, Crush, and Why Last Engaged Thread Fails," *Oil Gas J.*, **54**, 84–91 (Aug. 1, 1955).
2. Wais, J., Jr., "Recent Developments in Casing Standards and Design," Drilling and Production Practice, 249 (1947).
3. Strong, R. G. and Barkow, A. G., "Loopholes in the Pipeline Codes," *Oil Gas J.*, **55**, 162, 163, 165, 169, 171, 174, 176 (Sept. 16, 1957).
4. Shank, M. E., *Control of Steel Construction to Avoid Brittle Failure*, Welding Research Council, New York, N.Y. (1957).
5. McClure, G. M., Eiber, R. J., Hahn, G. T., Boulger, F. W., and Masobuchi, K., "Research on the Properties of Line Pipe," Summary Report Catalog No. 40/PR, 1962, American Gas Association, New York, N.Y.
6. Saylor, W. A., "Critical Materials and Substitute Steels Theme of ASM Show," *Iron Age*, **167**, 97–98 (Mar. 29, 1951).
7. Atterbury, J. T., "Stresses During Shipping, Handling and Laying Thin Wall Pipe," *Pipe Line News*, pp. 44–47 (Dec. 1962).

chapter 6

MATERIAL IDENTIFICATION

As service applications are becoming more severe, careful marking and identification of the materials involved are extremely important. Such identification should be applied initially by the mill producing the respective plate, pipe or tubing material.

Where proper identification is not applied directly to the material, any mix-up might then be caused by the warehouse, fabricator or at the erection site. At this stage determination of the responsibility may be quite difficult.

Failures resulting from incorrectly identified or used materials may occur during fabrication when incompatible materials are being joined, or they may occur during service where the incorrect material does not exhibit the required strength, corrosion resistance or other essential properties.

INCORRECT IDENTIFICATION

Base Metals

Incorrect identification of alloys has been a far more frequent cause of failure than might be expected. The danger of mix-up is particularly great when markings have not been applied directly to the surface of the alloy by die stamping or etching. However, even on die-stamped or etched products, a significant number of errors have occurred.

Material Identification

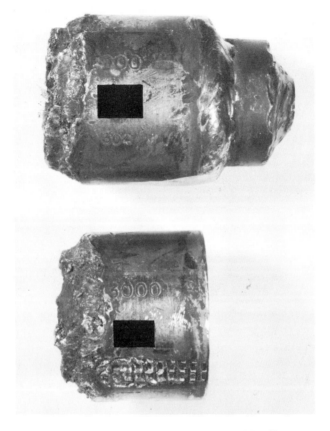

Fig. 6-1. Failure along weld of two 3000-lb couplings identified as Type 304 stainless steel. They were actually Monel. (Manufacturer's identification symbol has been blocked out.)

Figure 6-1 illustrates two 3000-lb couplings that failed after being welded to a Type 304 stainless steel header in a nuclear installation. The couplings had been purchased as Type 304 stainless steel. To avoid possible mix-up in the fabricator's shop, the manufacturer had applied the Type 304 markings and 3000-lb pressure rating directly to the surface of the couplings. Certified mill test reports also attested that the chemical composition conformed to ASTM requirements. The couplings failed in a service test. Subsequent examination revealed that the couplings actually were "Monel." Welding of

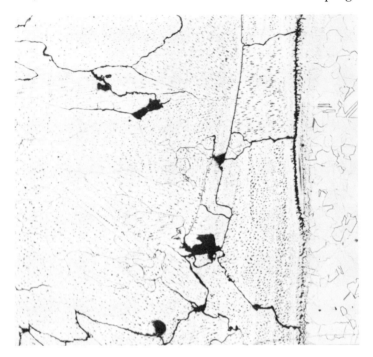

Fig. 6-2. Photomicrograph of weld metal (left) to base metal between Type 308 weld deposit and Monel (70 Ni–30 Cu) base metal illustrating "hot-short" cracking caused by copper penetration. (Magnification 100×)

"Monel" with stainless steel electrodes results in brittle joints as shown in Fig. 6-2, because of the copper content of the "Monel." It is fortunate that the pressure test revealed the defective joint. It might have held sufficiently during the brief test and then failed later in service. In a radioactive piping system, such a failure could render a plant inoperative for a long time.

Another example of material mix-up is illustrated in Fig. 6-3. The welding end cap shown, identified as Type 304 stainless steel, actually was Type 430.

Welding Filler Metals

There also have been several serious mix-ups by manufacturers of welding electrodes; one case involved a tool steel material identified

Material Identification

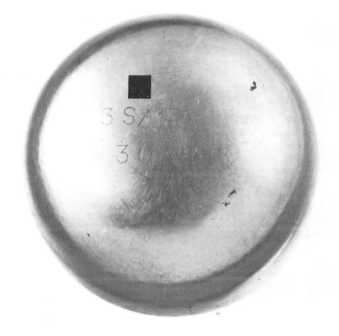

Fig. 6-3. Welding end steel cap identified as Type 304 stainless steel which was found to be Type 430 stainless steel. (Manufacturer's identification symbol has been blocked out.)

as stainless steel. It has also happened occasionally that an electrode or filler metal marked as one stainless grade is actually another.

The practice recently established by electrode producers to print the classifications directly on each electrode is a definite improvement, but may not necessarily avoid mix-up by manufacturers. Fortunately, instances of incorrect identification have been extremely isolated.

Absence of Identification

A number of service failures have occurred in steam power plants where, because of the absence of identification markings, carbon steel sections have been welded into critical high-temperature, high-pressure pipe lines, or into boiler tube systems requiring chromium-molybdenum alloy steels, or austenitic stainless steels. An example of a severe failure representing a sudden rupture in an incorrectly identified boiler tube is shown in Fig. 6-4.

Fig. 6-4. Failure in carbon steel boiler tube which has been identified as $2\frac{1}{4}$ Cr–1 Mo alloy steel material. The service involving temperatures at 1000°F required $2\frac{1}{4}$ Cr–1 Mo alloy steel.

A more gradual creep-type failure in a $2\frac{1}{8}$ in. OD by 0.300 in. wall boiler tube is shown in Fig. 6-5. Since the service temperature is 950°F, the tube material specified was a $1\frac{1}{4}$Cr–$\frac{1}{2}$Mo alloy steel. However, the material actually installed was carbon steel.

Because of material mix-ups and the use of carbon steel tubing where chromium–molybdenum alloy steel and stainless steel tubing were specified, the superheater tube sections suffering tube failures have had to be replaced completely in two major steam power plant boilers, even though the boilers had been erected in the early 1960's.

A failure which could have been very critical occurred during start-up testing of a nuclear submarine. The failure occurred in the 90 degree bend in a $1\frac{1}{2}$-in. Schedule 40 pipe. Subsequent examina-

Material Identification 93

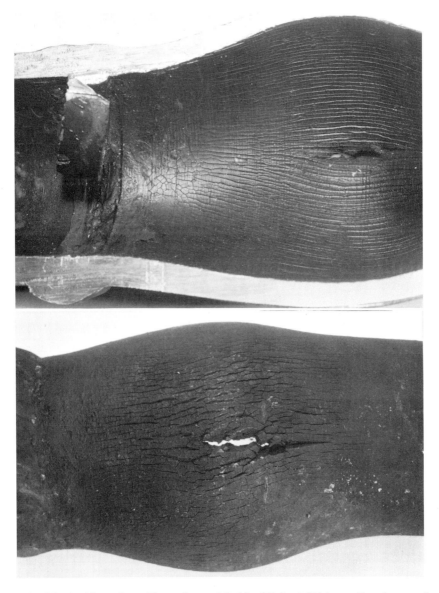

Fig. 6-5. Inside and outside surfaces of $2\frac{1}{8}$-in OD by 0.300-in. wall carbon steel tube section in service at 950°F, where $1\frac{1}{4}$ Cr–$\frac{1}{2}$ Mo alloy steel was required. (Magnification $1\frac{1}{4}\times$)

tion disclosed that the pipe was butt welded, and had been cold bent to a five diameter radius. The line of the butt weld was in the outer arc of the bead. Although the specifications called for seamless ASTM A106 Grade B pipe, the material actually installed was lap-welded A53 pipe. A106 pipe is fully deoxidized during the steel-making practice, whereas A53 pipe generally is not.

A similar mix-up in England resulted in a service rupture in an 8 in. steam main at 120 to 130 psi pressure after 6 years of operation. The rupture involved a 13-in. long split along a lap weld, resulting in violent escape of steam. Originally, seamless pipe had been specified for the piping system; however, 10 lengths were found to be lap-welded pipe of the type which had failed.[7]

Other examples of similar failures caused by mix-ups of materials have also been reported in Europe.[1-4] For example, failure by bulging has been reported in Germany in a steam power plant. The bulging was noted after 70,000 hours at 930°F in a 7-in. long spool piece of 6-in. diameter of $\frac{1}{2}$Mo steel pipe welded by mistake into a 1Cr–$\frac{1}{2}$Mo piping system. A larger pipe section of the $\frac{1}{2}$Mo pipe was also welded in the same piping system and failed by bulging and cracking involving an 11 to 15% increase in diameter in the cracked area.[1]

After approximately ten years of service, a severe rupture occurred in a 7-in. OD by 0.55-in. wall pipe section in a main steam piping system operating at 930°F and 1300 psi. Instead of 1Cr–$\frac{1}{2}$Mo alloy steel, a 10-ft long carbon steel pipe section had been used during a maintenance replacement operation.[2]

Some few manufacturers, because of ignorance, competition, or other reasons, upgrade their materials into a higher quality grade.

To protect themselves, the fabricator and user should insist on materials that bear proper identification markings, preferably applied by steel stamping.

There is no evidence of failures of materials due to die stamping, where the steel stamping has been done prior to hot forming or final heat treatment. Steel fittings for several decades have had Code and heat identifications applied by steel stamping prior to hot forming. No failures have resulted from this practice.

Codes are beginning to recognize the importance of identifications, preferably by steel stamping.

Where stress relieving is not done, the following is considered desirable: [5]

Stamping with low stress steel stamps is recommended for carbon and alloy steels containing up to 9% Cr.

Etching with an electric pencil is recommended for stainless steels and nonferrous materials.

CARELESS ERECTION PRACTICE

Sometimes, in spite of proper and permanent identification of the material type, the incorrect material is welded into a vessel or piping system.

A 4-in. diameter by 0.276-in. superheater tube failed after 7600 hours at 860 to 915°F, 1690 psi. The 1-in. long crack was preceded by bulging to a diameter of $4\frac{1}{2}$ in. The material specified was 1Cr–$\frac{1}{2}$Mo alloy steel. However, the material actually used was carbon steel. The carbon steel classification had been die stamped into the tube surface and was still evident after the failure.[6]

REFERENCES

1. Class, I., "Betriebserfahrungen mit Heissdampfleitungen im 500° Bereich," *Mitt. Ver. Grosskesselbesitzer*, **8**, 29–39 (1950).
2. Verein. d. Grosskesselbesitzer, 1955/56 Review.
3. Tichy, G., "Richtlinien für die Wärmebehandlung warmfesten Dampfkesselbaustähle unter Berücksichtigung örtlicher Erwärmung," *Mitt Ver. Grosskesselbesitzer*, **1**, 26–29 (1948).
4. Glockner, J., "Überhitzerschäden, reduzierende Atmosphäre, Werkstoffverwechslungen," *Mitt. Ver. Grosskesselbesitzer*, **11**, 184–188 (1950).
5. Pipe Fabrication Institute Standard ES-11, "Permanent Identification of Piping Materials," Sept. 1962.
6. Verein. d. Grosskesselbesitzer, 1958/59 Review.
7. Fuchs, E., "Quality Control Methods Adopted by a Large User When Purchasing Welded Equipment," *British Welding J.*, **6**, 429–438 (1959).

chapter 7

DEFECTS IN WROUGHT AND FORGED PRODUCTS

This chapter is concerned primarily with defects introduced by the manufacturer or processor of wrought or forged plate or piping materials. Such defects may have been introduced during the ingot stage, or they may have developed during the subsequent rolling, forming or reduction of the ingot or billet. Defects that result in service failures may also originate during the final plate rolling, the production of pipe in a pipe mill or forge shop or during the shaping of elbows, tees, flanges or other parts by a fitting manufacturer.

Some of these defects are difficult, if not impossible, to detect during manufacture and subsequent inspection. In critical service applications, they may then be responsible for failure of the pressure vessel or piping component.

NOTCH CONDITIONS

Failures from defects traceable to the original manufacturer usually result from one of the following:

(1) mechanical notches
(2) metallurgical notches

Mechanical notches involve actual discontinuities or separations in the material, either along the surface or on the inside.

Metallurgical notches pertain to differences in properties that are

Defects in Wrought and Forged Products

of sufficient magnitude to become potential causes of failure. They can occur within the same section of metal where localized heat treatment produces an area of high hardness adjacent to an area of lower hardness. Other typical metallurgical notches include hardened heat-affected zones within the plate or pipe material or adjacent to weld deposits. They represent also differences in properties as may occur between a wrought base metal and a weld deposit, which essentially represents a casting. More obvious metallurgical notches occur between dissimilar metals, as in weld joints between mild steel and stainless steel materials.

MECHANICAL NOTCHES

Laminations, laps, scabs, slivers, seams, and tears represent common types of metal separations in steel plate and pipe materials. They may be caused by defects present in the original ingot, or may have been caused during subsequent rolling or shaping of the plate or pipe material.[1]

Although different terms apply to these defects, they are often inadvertently referred to as just *laminations* or *seams*.

In most cases, these defects are not considered critical. They are not likely to affect the service characteristics of the steel since they usually occur parallel to the pipe surface. In fact, seam-welded plate pipe with laminations parallel to the surface can be compared to the layered vessel construction employed by some pressure vessel manufacturers. Laminations or seams that occur perpendicular or diagonal to the plate or pipe surface, however, should be considered as more critical notches. Some of these may completely transverse the wall thickness.

Common recognized defects are:

(1) Laminations
(2) Laps
(3) Slivers
(4) Scabs
(5) Seams
(6) Bark
(7) Cracks

98 *Defects and Failures in Pressure Vessels and Piping*

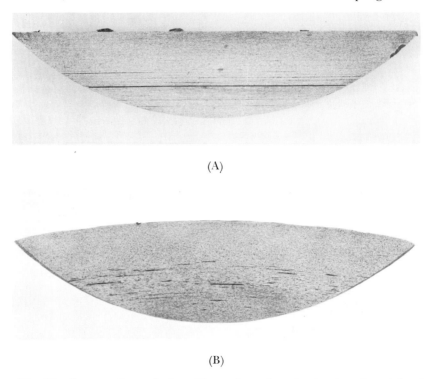

Fig. 7-1. Cross sections of plate illustrating minor laminations found longitudinally (A) and transverse (B) to the rolling direction. (Reduced to ¾×)

Laminations

Laminations normally occur at the inside of plate or pipe materials, and tend to be parallel to the surface. They are usually caused by inclusions or blow holes in the original ingot. Occasionally, a lamination is caused by a brick from the lining of the mold becoming entrapped within the ingot. During rolling, these inclusions become elongated and appear as fibrous longitudinal stringers of nonmetallics. Figure 7-1 illustrates cross sections of such stringers in the longitudinal and transverse direction of a plate material. Such narrow stringers generally would not be considered dangerous or rejectable.

Figure 7-2 illustrates a more severe lamination extending in length and width for several inches. It was originally extended during plate (cross) rolling, prior to forming and electric resistance welding into

Defects in Wrought and Forged Products 99

Fig. 7-2. Severe lamination in pipe considered rejectable.

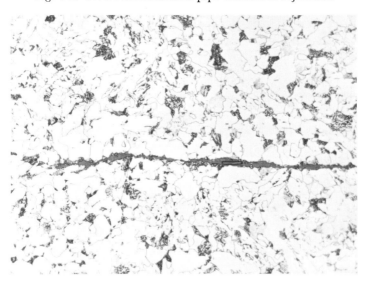

Fig. 7-3. Minor lamination in ½ Cr–½ Mo seamless pipe material which caused cracking failure during hot extrusion. (Magnification 250×)

100 *Defects and Failures in Pressure Vessels and Piping*

Fig. 7-4. Failure during extrusion operation of 6-in. diameter outlet caused by minor lamination shown in Fig. 7-3. (Header 1/6 size; detail 1/3 size)

pipe. This degree of severity should, of course, be a cause of rejection of the steel. Since the lamination did not extend to the plate edge, it was not detected until a circumferential cut was made for pipe shop welding.

Heavy laminations and, sometimes even minor laminations, may cause cracking or failures during final fabrication. They may also result in defects or cracking in weld deposits, Figs. 13-24 and 13-25.

A minor lamination which caused failure during the extrusion of an outlet in a header is illustrated in Fig. 7-3. The split in the 6-in. diameter outlet in the 12-in. diameter pipe is shown in Fig. 7-4.

Laps

Laps, illustrated in Fig. 7-5, represent another common defect caused during plate or pipe rolling, as may be produced in a pilger mill. It involves rolling of metal protrusions left from previous passes into the surface during the next pass.

Such defects generally appear parallel to the pipe surface and, if not deep, are not considered dangerous or rejectable. The lap illustrated in Fig. 7-5 had a depth of less than 2% of the wall thickness.

A more severe lap is shown in Fig. 7-6. This lap caused rejection

Defects in Wrought and Forged Products

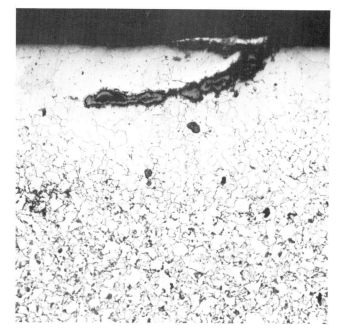

Fig. 7-5. Example of minor surface lap commonly found on many commercial plate products. This type is not usually considered detrimental. (Magnification 100×)

of the 3-in. nominal OD Type 304 stainless steel pipe, since its length exceeded the 25% permissible length limit of the applicable specification (ASTM A405).

Laps of minor depth, and considered acceptable under applicable standards may nevertheless cause leaks in vessels, headers and pipe. Sometimes, such laps are not even detected unless nozzle or other openings are cut into the material. Where they cut through surface laps, liquid or gas may seep from the nozzle to the surface. In such cases, the leakage can frequently be detected by a soap bubble test as illustrated in Fig. 7-7. In a few instances, however, such a test, as well as a subsequent hydrostatic test, has not revealed the presence of surface laps. Failures, evidenced by leaks, may then occur after years of service, often involving opening of the lap and some extension by cracking.

102   *Defects and Failures in Pressure Vessels and Piping*

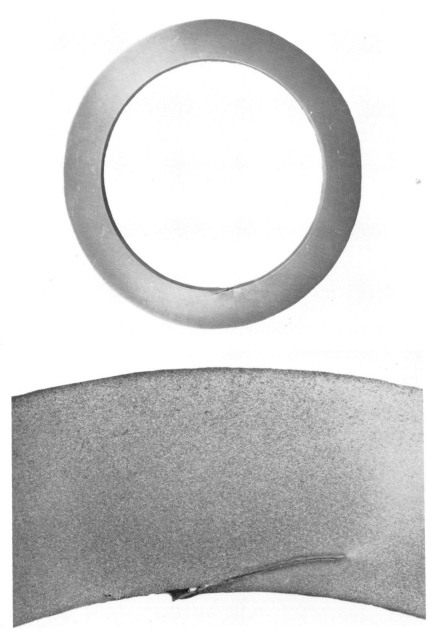

Fig. 7-6. A rejectable lap defect on the inside of a Type 304 stainless steel pipe section for nuclear plant service. (Magnification 5×)

Defects in Wrought and Forged Products 103

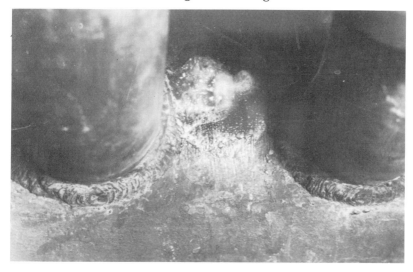

Fig. 7-7. Leak detected in soap bubble test caused by surface lap reaching nozzle outlet hole drilled through shell of header.

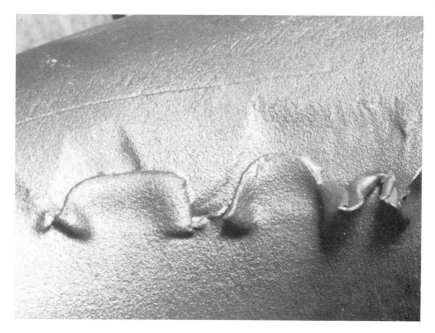

Fig. 7-8. Severe surface laps (overfill) on pipe which opened up during hot bending.

104   *Defects and Failures in Pressure Vessels and Piping*

In seamless tubing, laps are also referred to as *overfill* or *fin,* Fig. 7-8. These defects are produced during the sizing operations after piercing. Metal is squeezed out between the two rolls of the high mill and is subsequently rolled down and pressed into the outside tube surface during the second pass. In severe cases, overfill defects have reduced the bursting strength significantly. For example, a 7-in. OD, 26-lb N-80 oil-well casing burst at 1500 psi instead of 6600 psi due to an overfill.[2]

An example of a major failure caused by a lap defect occurred in a carbon steel boiler feed pipe 6¾-in. OD by 0.28-in. wall carrying

Fig. 7-9. Example of surface sliver rolled into pipe surface.

Defects in Wrought and Forged Products

water at 390°F and 1470 psi. After 22,000 hours, a slight leak was observed. To avoid interruption of the operation, the leak was closed by a tube strip liner applied over the defect. Three hours later, a 5-ft long rupture occurred in the pipe causing two employees to receive severe burns from the hot water. Subsequent examination revealed severe lap defects in the pipe steel.[3]

Slivers

Slivers are metal surface ruptures, usually very thin and rolled into the surface. Slivers may also involve small sections of scale embedded in the surface during rolling. These defects are also referred to as *tongues*. When considered undesirable, they can be readily removed by grinding.

A more severe sliver rolled into a pipe surface is illustrated in Fig. 7-9.

Scabs

Scabs represent scale rolled into the surface. This defect is generally very shallow and is not usually considered objectionable. An example of a rolled-in scab in an elbow surface which became dislodged during subsequent cleaning is shown in Fig. 7-10.

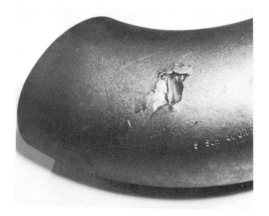

Fig. 7-10. Example of scab defect in steel elbow which became dislodged during subsequent cleaning.

106 *Defects and Failures in Pressure Vessels and Piping*

Fig. 7-11. Minor and severe seams in pipe.

Defects in Wrought and Forged Products

Fig. 7-12. Severe seam defect through pipe wall which opened up during hot bending.

Seams

Seams are the result of hot shortness, blow holes, or inclusions which have not been welded together during rolling. They tend to be elongated in the direction of rolling and often appear as shallow, narrow parallel fissures on the surface of the plate. On pierced tubing and pipe seams, they may appear spirally around the pipe, Fig. 7-11. Generally, seams are not harmful unless deep. Some seams are very tight and appear as hairline indications. Occasionally, they can be quite deep and should be explored by grinding. Some deep seams may pass completely through the wall thickness. They may be sufficiently tight and not cause leaks in a hydrostatic test, unless high pressures are involved. Seams across the wall may also open up during service as a result of fatigue, steam pressure, etc.

Severe seams through the pipe wall may also result in ruptures during bending and other forming operations, Fig. 7-12. When deep, they can act as crack starters and lead to failure, particularly under internal pressure.[2] Seams have been responsible, also, for cracking and failures in boiler tubing[4] and in main steam piping—9½-in. OD by 0.59-in. wall operating at 930°F, 1000 psi.[5]

Bark

Bark represents intergranular penetrations of oxides and scale into the surface of tubing, usually on the inside, Fig. 7-13. They are usually not considered harmful.[6]

108 *Defects and Failures in Pressure Vessels and Piping*

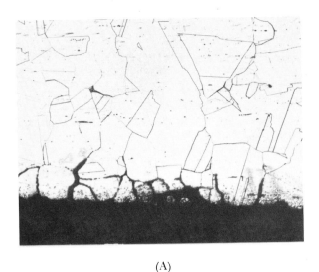

(A)

Fig. 7-13. Examples of oxide penetration referred to as "Bark" on the inside of stainless steel tubing. (A) 1⅝-in. OD by 1⅛-in. wall type 316—unetched and etched. (Magnification 100×) (B) 1⅛-in. OD by ⅜-in. wall Type 347. (Magnification 250×)

Defects in Wrought and Forged Products 109

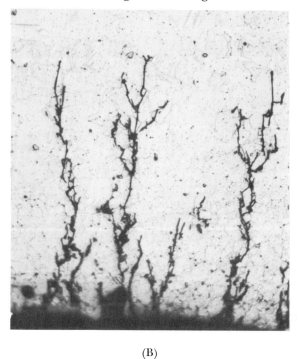

(B)

Fig. 7-13 *cont'd*

Cracks

Occasionally, cracking occurs during plate rolling or pipe forming. Minor cracking in a plate is shown in Fig. 7-14. More severe conditions of cracking are also occasionally found in commercial plate and pipe materials.

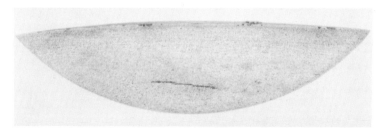

Fig. 7-14. Cross section illustrating minor cracking in 1 Cr–½ Mo plate material.

110   *Defects and Failures in Pressure Vessels and Piping*

(A)

(B)

Fig. 7-15. Example of lamination in neck of forged flange. (Magnifications (B) 5×, (C) 50×)

Defects in Wrought and Forged Products

(C)

Fig. 7-15 cont'd

Code Acceptability Limits

Many specifications covering plate and pipe materials contain acceptability limits of surface defects.

For example, ASTM Specification A335,[7] which covers seamless ferritic alloy steel pipe containing up to 9Cr and 1Mo, states:

"Pipe showing scabs, seams, laps, tears, or slivers not deeper than 5% of the nominal wall thickness need not have these defects removed. If deeper than 5%, such defects shall be removed by machining or grinding.

"Pipe showing inside or outside surface checks, (fish scale) $1/64$ in. or less in depth need not have these defects removed. Such defects over $1/64$ in. but not more than $1/32$ in. in depth shall be removed by

112 *Defects and Failures in Pressure Vessels and Piping*

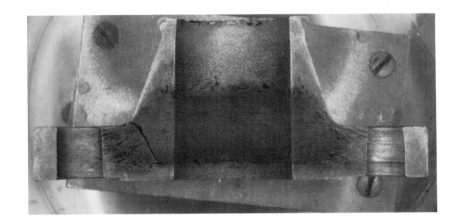

Fig. 7-16. Examples of cracks in forged steel flanges.

machining or grinding. Pipe on which these defects are more than $\frac{1}{32}$ in. in depth shall be rejected, unless the manufacturer can demonstrate to the purchaser that the defects are not injurious."

Injurious defects are defined as follows:

"All defects shall be explored for depth. When the depth is in excess of $12\frac{1}{2}\%$ of the nominal wall thickness or encroaches on the minimum wall thickness, such defects shall be considered injurious."

Forging Defects

Many components used in pressure vessels and piping are produced by forging methods. Inherently defective steel billets or improper forging methods may result in defective products.

Examples of the more common defects encountered occasionally in forged products such as in flanges are laminations, Fig. 7-15, and cracks, Fig. 7-16.

Similar defects in forged and bored carbon steel tube "Y" fittings 2.75-in. OD by 0.224-in. wall have resulted in leaks, particularly along the circumferential butt welds between the fittings and tubing.[3] The leaks were detected during the initial hydrostatic test.

Bar Stock Defects

Defects in bar stock can also result in defects and failures.

During the initial steam-up of a new boiler, leaks developed in a flow nozzle section in the 14-in. OD by 1.9-in. wall main steam piping. Examination showed that the four holding pins had cracks in their centers, as illustrated in Fig. 7-17. These cracks appear to have been present in the original bar stock as a result of defects in the billet which had not been cropped sufficiently. The cracks were then extended during subsequent rolling and heat treatment. The cracks were sufficiently tight so as not to leak during the hydrostatic pressure test. Only the final steam start-up test revealed their presence.

Tool Marks and Gouges

Mechanical notches produced by plate rolling or by pipe or tube drawing or rolling, by seam welding of pipe or fittings, or by machining and other production operations can also have a detrimental effect on the service life of the fabricated material.

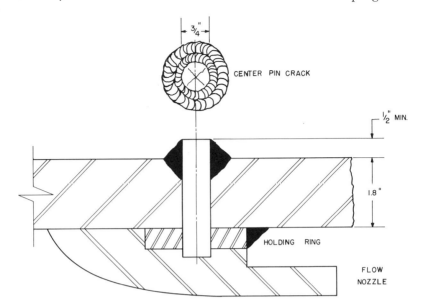

Fig. 7-17. Defects in bar stock causing leaks in $2\frac{1}{4}$ Cr–1 Mo piping system.

Figure 7-18 illustrates a tube failure that originated at a plug score groove formed on the inside of the tube by the drawing operation. These grooves are normally caused by hard pieces of metal which adhere to the high-mill plug during the tube rolling operation which follows piercing. Generally, the bottom of plug scores appears round so that they are of little consequence. Occasionally, they exhibit a sharp notch. Mechanical vibrations initiate and propagate cracking at the notch formed by the die groove, as shown in Fig. 7-18. Environments involving stress corrosion are even more likely to initiate cracking in die grooves. This is shown in Fig. 7-19.

Plug scores have also caused failure in oil well tubing due to fracturing along the groove.[2] Unfortunately, normal measurements of their depth has not been a rejectability criterion, as it does not measure the severity of the notch. Allowance tolerances of the American Petroleum Institute permit defects to maximum of $12\frac{1}{2}\%$ of the wall thickness. Some plug scores significantly deeper than $12\frac{1}{2}\%$ of the wall thickness have not resulted in failure, while other grooves shallower than $12\frac{1}{2}\%$ but with sharp notches have caused failures.[2]

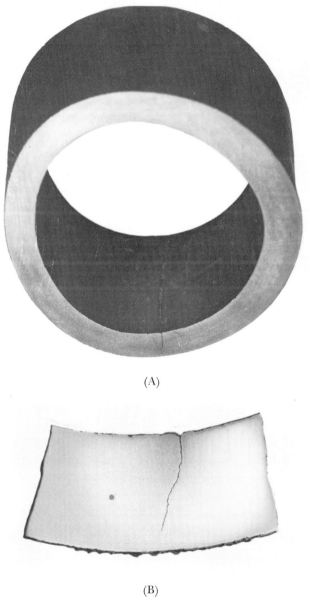

(A)

(B)

Fig. 7-18. Failure in tube caused by notch formed along tube inside by tube drawing die. (A) Cross section of tube (B) Cross section across tube wall containing crack. (Magnification 5×)

116   *Defects and Failures in Pressure Vessels and Piping*

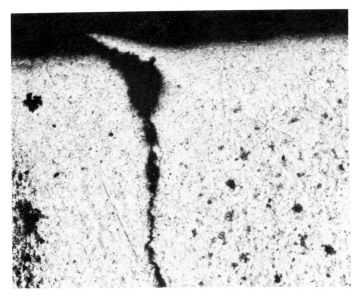

Fig. 7-19. Stress-corrosion cracking starting from drawing die tear in boiler tube. (Magnification 50×)

Gouges have also been responsible for a number of failures in 16-, 20-, and 26-in. gas line pipe.[8,9] Some of these failures occurred during pressure testing. Metallographic examination showed a localized region of highly cold-worked steel underneath the gouge.[9] In several instances, a layer of untempered martensite was present on the pipe surface in the gouge. This untempered martensite is a high-strength, hard and brittle constituent susceptible to cracking (acting as a localized metallurgical notch—see below). The failure was believed to be caused by cracking in the martensite layer when the gouge was made, or later at a low internal pressure. Propagation across the wall thickness occurred as a result of further pressurizing.

Notches in Welded Pipe Seams

Figure 7-20 illustrates a notch condition occasionally encountered in welded piping, such as those of ASTM A53 materials. From the smooth outside surface, one would almost assume that the material were seamless. There have been a number of instances where failures, originating at notches present at the root of the weld, have occurred in seam-welded piping materials.

Defects in Wrought and Forged Products

Fig. 7-20. Cross section of continuously welded ASTM A-53 open-hearth steel pipe with notch and cracking along inside of pipe. (Magnification 7½×)

METALLURGICAL NOTCHES

Hot Shortness

Definition. Hot shortness defined as "brittleness in hot-metal" is generally related to films of impurities and nonmetallic materials which are present in grain boundaries and, in certain temperature ranges, reduce significantly the cohesive strength (hot ductility) of the material. Although some austenitic stainless steels may become hot short in the temperature range between 1400 and 1800°F, hot shortness occurs more readily at higher temperatures, particularly near the melting point of the metal. Impurities, nonmetallic inclusions and metal phases concentrated along the grain boundaries melt at temperatures below the melting point of the alloy. This is also referred to as *liquation*. Type 347 stainless steels (see Chapter 4) and copper bearing steels [10] represent typical examples of susceptible steels.

An example of hot shortness in a copper-bearing stainless steel is shown in Fig. 7-21. Grain boundary liquation of the high copper phase occurred at about 2400°F. At lower temperatures, liquation tends to be less pronounced. Structural phases of the "eutectic" type pattern illustrated in Fig. 7-22, nevertheless, are evidence of a hot short condition at about 2200°F.

However, the presence of low melting phases or other brittle grain

118   *Defects and Failures in Pressure Vessels and Piping*

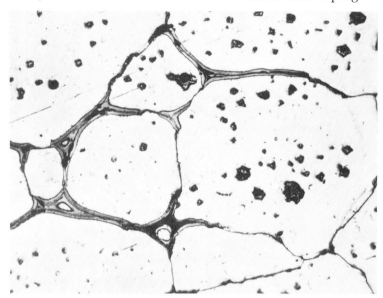

Fig. 7-21. Example of severe grain boundary liquation in copper-bearing stainless steel. (Magnification 250×)

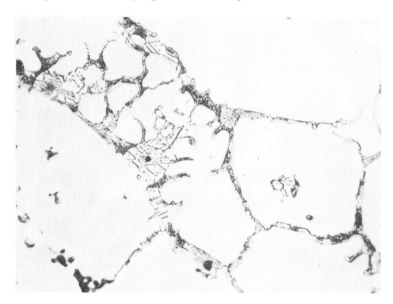

Fig. 7-22. Severe degree of hot shortness along grain boundaries of copper bearing stainless steel illustrating "Chinese Script" type eutectic structure. (Magnification 500×)

boundary film may be very minute and detectable only with difficulty by microscopic techniques. Figure 7-23 shows photomicrographs illustrating fine hot cracking and a fine grain boundary phase in a forged vessel dome made of Type 347 stainless steel. These defects became apparent only when mass spectrometer testing indicated very low leak rates.

Hot shortness may be introduced externally or it may be an unexpected or known characteristic of a particular alloy. Some alloys are inherently hot short and must be worked, fabricated or welded with extreme care.

External Causes. When steel plate and pipe materials heated to hot working or hot forming temperatures come in contact with nonferrous metals such as copper, lead, zinc, or tin, the surface of the steel will become embrittled because of penetration of these metals into the steel. This condition is also referred to as hot shortness.

Figure 7-24 is a photomicrograph illustrating copper penetration into a $2\frac{1}{4}$Cr–1Mo pipe, particularly along the grain boundaries of the steel. Hot bending or forming of the affected areas would readily result in cracking. In severe instances of penetration, cracks can develop and propagate in service involving thermal fatigue or stress corrosion.

Similar conditions hold true also for the nonferrous metals. For example, nickel alloys are severely embrittled by sulfur.

It cannot be sufficiently emphasized that it is extremely important that plate, pipe, fittings, and similar materials be carefully handled during their processing from the ingot to the final shape.

Inherent Characteristics. Hot shortness may also be an inherent characteristic of a metal. In some alloys, hot shortness may be completely unexpected. In other materials, hot shortness may be an expected characteristic when it is caused by specific alloying elements, purposely present to provide certain properties such as strength or corrosion resistance.

Unexpected hot shortness can usually be related to improper control over chemistry. It may also be caused by minor additions of certain alloying elements to provide improved fluidity in steel making (pouring) or casting practice.

Improper control of the chemistry of steel which has made a steel inherently hot short may result in the steel becoming susceptible to

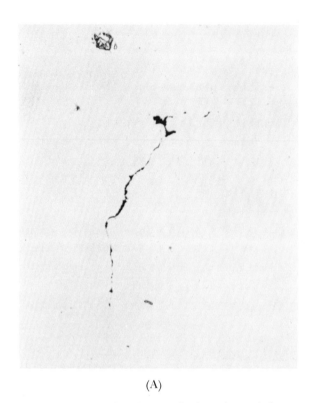

(A)

Fig. 7-23. Microfissuring (1) in forged vessel dome caused by hot-short condition and grain boundary films in Type 347 stainless steel. (A) unetched, (B) etched. (Magnifications 500×)

cracking during hot forming. Figure 7-25 illustrates severe cracking of a 30-in. diameter 1⅛-in. thick pipe. The cracking occurred during hot bending by approximately 15 degrees on a 20-ft radius at 1900°F. The pipe was produced from wrought 1¼Cr–½Mo plate steel material that had been seam welded by the submerged-arc welding process. Subsequent hot tensile tests of the base metal shown in Fig. 7-26 confirmed that the steel was hot short and did not contain sufficient ductility at normal hot bending temperatures.

For a pipe material to be suitable for hot bending to a five diameter radius, tensile test specimens should exhibit a ductility of at least 20% at the temperatures involved in hot bending.

Unexpected inherent hot shortness may also be responsible for

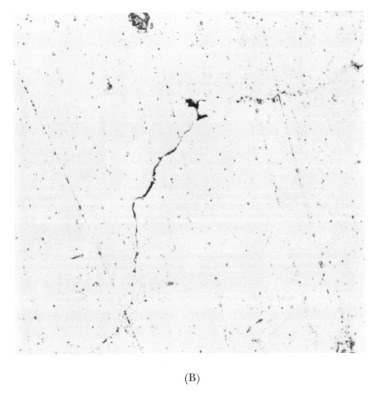

(B)

Fig. 7-23 cont'd

cracking during welding occurring in the heat-affected zone adjacent to the weld deposit. This base-metal cracking is further discussed in Chapter 12.

Some plate and pipe materials are inherently hot short and susceptible to cracking when hot working or when welding. Copper-bearing aluminum alloys are typical examples.[11] Figure 7-27 illustrates cracking in the heat-affected zone and adjacent weld area of 2024-T6 aluminum pipe. The heat generated by welding causes remelting of the low melting (eutectic) constituents in the alloy, primarily at the grain boundaries. This so-called intermetallic phase is brittle. Ordinary weld stresses tend to cause fissuring in the heat-affected zone, Fig. 7-28. Even buttering (or overlaying) the surface of the plate or pipe bevel with pure aluminum, such as 1100, gen-

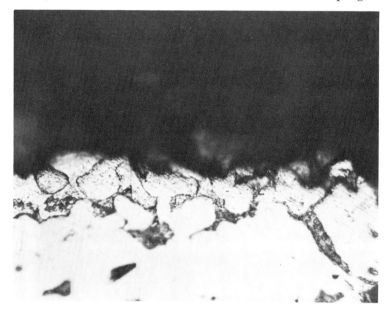

Fig. 7-24. Copper penetration into surface of 2½ Cr–1 Mo pipe caused by contact of red hot pipe with copper bearing material. (Magnification 500×)

Fig. 7-25. Severe cracking in 30-in. OD by 1⅛-in. wall 1¼ Cr–½ Mo pipe during hot bending.

erally will not eliminate this cracking problem.[12] Other copper bearing and magnesium-zinc [13] bearing heat-treatable aluminum alloys susceptible to this heat-affected zone and weld fissuring in the base and weld metal are 2014, 2219, 7075, and 7079.

In copper and copper-nickel bearing steel plate and pipe materials, hot forming or high temperature heat treatments may cause enrichment of copper and precipitation of CuNi phases along the grain boundaries. During service, this condition may lead to surface cracking. Room temperature hydrostatic pressure tests of steam drums (0.2C, 0.9Mn, 1.1Ni, 1.1Cu) made of CuNi alloy steels failed in a brittle manner when improper heating resulted in the formation of a copper-rich phase along the grain boundaries. Even bend test specimens exhibited zero ductility.[14] Heat treatment for 2 hours at 1690°F and 4 hours at 1110°F may restore some bend ductility.[14]

Surface Carburization and Decarburization

Surface metallurgy is rarely considered in evaluating plate, pipe or similar materials. Nevertheless, carburization or decarburization may critically affect the welding and service characteristics of materials.

During hot forming into plate, pipe, fittings, etc., care should be exercised to insure that excessive carburization or decarburization of the steel surface does not occur.

For example, in the forming of steel elbows by conventional techniques, tubular sections are heated by means of gas burners to temperatures on the order of 1650 to 1850°F. The sections are subsequently "pushed" over special mandrels to form the elbows. If the gas used to heat the tubular sections is not "neutral" but instead has a composition that tends to be carburizing, the steel surface may become severely carburized. This may occur also in the gas heating of flat steel shapes for hot forming into half elbows and tee sections subsequently welded together and marketed as welded fittings. Gas heating for pipe bending and straightening and many other hot forming and shaping operations may produce similar results.

Figure 7-29 illustrates surface carburization in a seamless elbow of $1\frac{1}{4}$Cr–$\frac{1}{2}$Mo low-alloy steel composition, such as is widely used in critical high-pressure high-temperature piping systems. Butt welds were subsequently made on this material; they included proper pre-

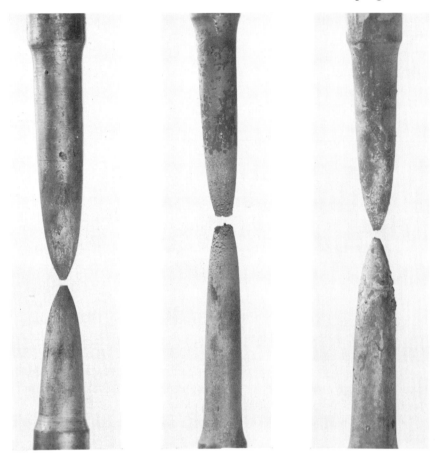

Normal 1¼ Cr–½ Mo Grade.

Fig. 7-26. Hot tensile tests of normal steel (top three specimens) are compared with those of hot short steel (three specimens on facing page). Testing temperatures for both normal and hot short steel were, from left to right, 1400°F, 1600°F, and 1850°F. Elongation in 2-in. for normal steel was, from left to right, 56%, 65%, and 75%. Elongation in 2-in. for hot-short steel was, from left to right, 43%, 20%, and 5%.

heating to 500°F and stress relieving at 1325°F. These welds failed during bend tests upon bending by as little as 45 degrees. This is shown in Fig. 7-30. Had a lug or guide plate for hanger attachments been welded to this fitting or a nozzle connection been made, failure might have resulted in service from severe stresses caused by pipe

Defects in Wrought and Forged Products

Hot Short 1¼ Cr–½ Mo Grade.

Fig. 7-26 cont'd

movement. If such a failure were to occur in service, it is likely that the fabricator who made the attachment weld, rather than the fitting manufacturer would have been held responsible. Even if a chemical analysis certificate had been furnished, the surface carburization would not have been discovered. Only weldability tests and possibly photomicrographs would have revealed this condition. To require such tests on all materials furnished would have been prohibitive in cost.

126   *Defects and Failures in Pressure Vessels and Piping*

Fig. 7-27. Photomicrograph illustrating fissuring along the grain boundaries of hot-short 2024-T6 aluminum pipe.

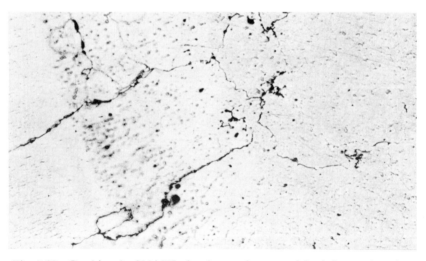

Fig. 7-28. Cracking in 2024-T6 aluminum pipe caused by inherent hot shortness of pipe base metal. (Magnification 100×)

Defects in Wrought and Forged Products 127

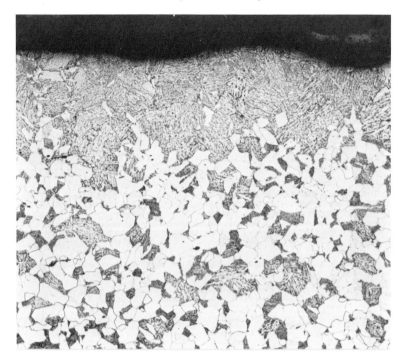

Fig. 7-29. Surface carburization revealed by metallographic examination of microstructure in $1\frac{1}{4}$ Cr–$\frac{1}{2}$ Mo low-alloy steel fitting. (Magnification 100×)

Fig. 7-30. Low ductility along weld seam causing cracking in bend test of coupon removed after stress-relief heat treatment from carburized fitting.

Hardness Variations

The hardness of a steel varies primarily with its chemical composition and the heat treatments given. A higher hardness tends to increase the yield and tensile strengths and reduce ductility.

If the difference in hardness exceeds approximately 70 to 100 points Brinell and thermal and/or mechanical fatigue are involved in the service, failures may result.

A number of failures have occurred in high-strength API 5LX-52 gas-transmission pipe starting in areas containing local hard spots.[9] The hardness in these spots ranged from 301 to 372 Brinell, whereas the nominal pipe hardness was in the range from 182 to 228 Brinell. The hard spots were approximately 3 to 6 in. in diameter. They were believed to have originated during the rolling of the plate prior to forming into pipe and seam welding. Although the exact cause could not be determined, it was considered possible that during rolling of the plate, the line was stopped and cooling water from the rolls produced localized quenching of the skelp and formed the hardened area.

A high hardness area caused by local water quenching in a high-carbon steel oil well casing resulted in a cracking failure in the body of the pipe 1300 ft below ground level after two months of service.[15]

REFERENCES

1. Thompson, H., "Steel Defects and Their Detection," Pitman and Sons, London, 1952.
2. Texter, H. G., "Why Oil-Well Tubing and Casing Fail. Why They Burst, Leak, Crush, and Why Last Engaged Thread Fails," *Oil Gas J.*, **54**, 84–91 (Aug. 1, 1955).
3. Verein. d. Grosskesselbesitzer, 1957/58 Review.
4. Verein. d. Grosskesselbesitzer, 1959/60 Review.
5. Verein. d. Grosskesselbesitzer, 1955/56 Review.
6. Hoke, J. and Eberle, F., "Experimental Superheater for Steam at 2000 psi and 1250°F.—Report After 14,281 Hours of Operation," *Trans. ASME*, **79**, 307–317 (1957).

7. ASTM Specification A335-63T, "Seamless Ferritic Alloy Steel Pipe for High-Temperature Service," ASTM Standards, **1**, American Society for Testing and Materials, Philadelphia (1964).
8. Saylor, W. A., Comments reviewed in *Iron Age,* **167**, 97–98 (Mar. 29, 1951).
9. McClure, G. M., Eiber, R. J., Hahn, G. T., Boulger, F. W., and Masubuchi, K., "Research on the Properties of Line Pipe," Summary Report Catalog No. 40/PR, 1962, Am. Gas Assoc., New York.
10. Thielsch, H., "Copper in Stainless Steels," Welding Research Council, Bulletin No. 9 (Aug. 1951).
11. Houldcroft, P. T., "The Welding of High-Strength Heat-Treated Aluminum Alloys," *British Welding J.,* **5,** 261–271 (1958).
12. Phillips, A. L., "The Welding Handbook," Section 1, 5th Ed., pp. 4.107–4.111, New York, The American Welding Society, 1963.
13. Mudrack, K. P., "Schweisstechnische Untersuchungen einer Al-Zn-Mg Legierung unter besonderer Berücksichtigung der Schweissrissigkeit," *Schweissen und Schneiden,* **12,** 45–55 (1960).
14. Wellinger, K., "Die Bedeutung von Stählen mit höherer Streckgrenze im Kesselbau," *Mitt. Ver. Grosskesselbesitzer,* **91,** 288–297 (1964).
15. Texter, H. G., "Why Oil-Well Tubing and Casing Fail in Tension, and Why They Collapse," *Oil Gas J.,* **54,** 86–96 (July 4, 1955).

———chapter 8

CASTING DEFECTS

As holds true for wrought and forged pressure vessel or piping materials, perfect castings do not exist.

Defects may have been present in the original casting, or may have been produced in the subsequent processing or even repair of the casting.

Failures can arise from (1) mechanical notches or from (2) metallurgical notches. (See Chapter 7.)

MECHANICAL NOTCHES

Metal separations are particularly common in static castings. In fact, a more liberal interpretation is generally applied to defect limitations on radiographs of castings than of weld deposits—even though the cast components and welds may be located in the same pressure vessel or piping system. For example, larger and more severe groupings of gas pockets are acceptable in cast valve bodies [1] than would be permissible in weld deposits.[2, 3]

Common recognized casting defects are:

(1) Gas and blow holes
(2) Sand spots
(3) Inclusions
(4) Internal shrinkage

Casting Defects

(5) Hot tears and cracking
(6) Unfused chaplets
(7) Internal chills

Gas and Blow Holes

Gas and blow holes in castings may range from fine porosity to large cavities. They are caused by gases dissolved in the molten metal, which are evolved and entrapped during solidification. Although generally round, blow holes may also be oval or elongated, or have irreguar shapes.

Gas and blow holes on the inside of castings rarely have resulted in service failures. In fact, there has been no record of failures caused by internal gas pockets within the most liberal acceptability limit of ASTM Specification E71, Fig. 8-1.[1]

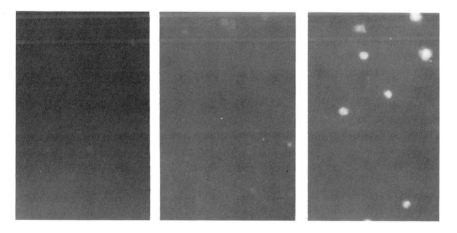

Fig. 8-1. Prints of radiographs illustrating Class 1, 2, and 5 acceptability limits in ASTM Specification E71 governing the most severe and most liberal limits.[1]

Inclusions

Inclusions represent scale, oxides or dross entrapped within the casting during the process of solidification. Such inclusions frequently appear in a dendritic pattern, as illustrated in Fig. 8-2.

These defects, similar to slag inclusion in welds, generally are not considered critical unless they exceed applicable specification limits, or occur in significant size at or near the surface.

132 *Defects and Failures in Pressure Vessels and Piping*

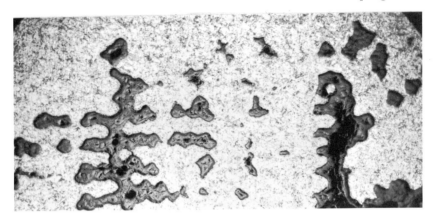

Fig. 8-2. Inclusion in cast ½ Mo steel valve 12-in. OD by 1-in. wall. (Magnification 25×)

Shrinkage Cavities

Shrinkage cavities in steel castings represent a common defect.[4] Large shrinkage cavities result from improper gating, risers of insufficient size, or from the failure to locate risers over heavy sections or hot spots in the casting. Smaller cavities may be due to an insufficient number and location of risers which exceeds the normal feeding distance of risers.

Shrinkage cavities generally tend to be most prevalent toward the center of the casting, where solidification occurs last because hot molten metal was not available to fill the void created by the volume change from liquid to solid.

Minor shrinkage cavities in cast low alloy and stainless steel valves are illustrated in Fig. 8-3. A photomicrograph illustrating the appearance of minor shrinkage cavities is shown in Fig. 8-4. A major shrinkage cavity is shown in Fig. 8-5. The cavity measured approximately 1½ in. by 6 in. by ¼ in.

Very severe shrinkage is illustrated in Fig. 8-6. This amount is obviously rejectable.

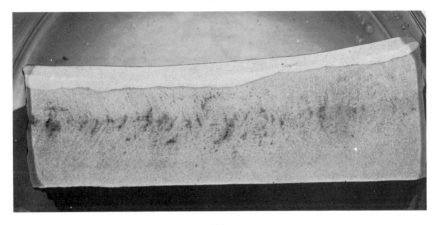

(A)

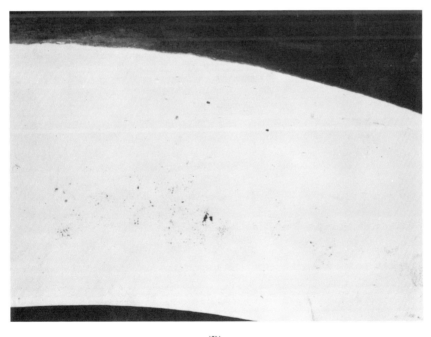

(B)

Fig. 8-3. Examples of cross sections illustrating minor shrinkage cavities in cast valve bodies. (A) Cast $1\frac{1}{4}$ Cr–$\frac{1}{2}$ Mo steel valve body with shrinkage and a weld repair on outside surface $1\frac{1}{2}$ in. thick (reduced to $\frac{3}{4}\times$). (B) Type 304 stainless steel valve body 0.300 in. thick. (Magnification $9\times$)

134   *Defects and Failures in Pressure Vessels and Piping*

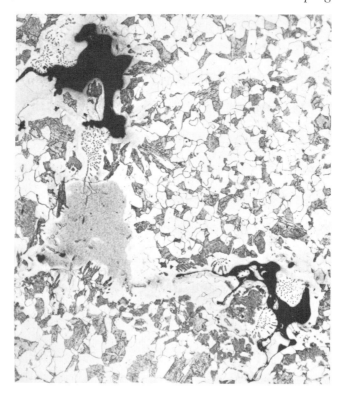

Fig. 8-4. Photomicrograph illustrating typical minor shrinkage cavities in ½ Mo steel casting. (Magnification 100×)

Fig. 8-5. A major shrinkage cavity in cast ½ Mo steel valve approximately 1⅜ in. thick.

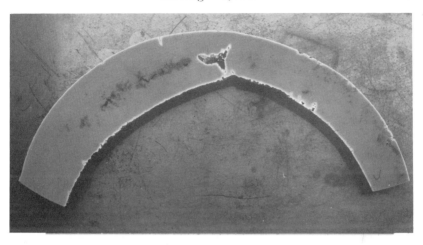

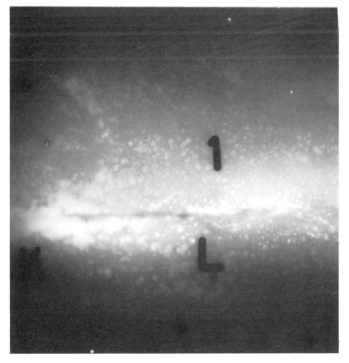

Fig. 8-6. Cross section and radiograph show severe shrinkage defect and hot tears in 5 Cr–½ Mo casting furnished to ASTM Specification A217.

Test Results

Tensile tests of radiographically sound castings having a tensile strength of 114,000 psi, lost 4.5% strength to 109,000 psi for sections of Class 2, ASTM E71 shrinkage porosity (see Fig. 8-1). At a shrinkage level of Class 5 to 6, the strength decreased 8% to 105,000 psi.[5]

Service Experience

Shrinkage cavities have resulted in a number of service failures. Figure 8-7 shows cracking in a $1\frac{1}{4}$Cr–$\frac{1}{2}$Mo valve body in a main steam piping system at 950°F. Cracking started from shrinkage cavities, as illustrated in Fig. 8-8, from the section removed from the valve wall.

Fig. 8-7. Cracking across wall of 12-in. OD cast $1\frac{1}{4}$ Cr–$\frac{1}{2}$ Mo alloy steel valve in main steam piping system.

Casting Defects 137

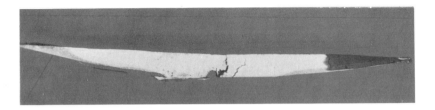

Fig. 8-8. Cross section of section removed from valve which failed by cracking and illustrating shrinkage cavities. (Magnifications 1× and 6×)

Hot Tears and Cracks

Hot tears and cracks are very similar in appearance. As the name implies, hot tears would occur during the initial stages of cooling. Cracking is normally considered to occur at or near room temperature. Compositions susceptible to hot shortness may develop hot tears in the as-cast condition. Hot tears often tend to contain oxide or scale, whereas cracks may be scale-free.

Hot tears at the surface of a 5Cr–½Mo steel casting are illustrated in Fig. 8-9. The steel casting was produced to ASTM Specification A217.

On radiographic films, internal shrinkage cavities can present a problem in interpretation and may, at times, be difficult to distinguish from hot tears or cracks. The latter types of casting defects are not considered acceptable for pressure vessel and piping applications.

138 *Defects and Failures in Pressure Vessels and Piping*

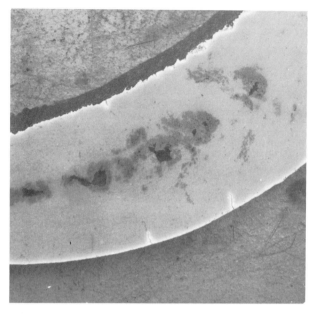

Fig. 8-9. Hot tears at the surface of a 5 Cr–$\frac{1}{2}$ Mo steel casting.

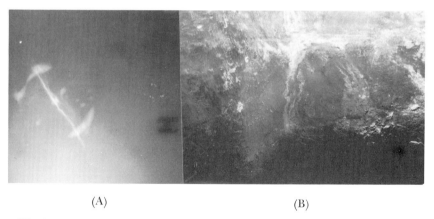

(A) (B)

Fig. 8-10. Cracking and leak caused by unfused chaplet in $\frac{1}{2}$ Mo steel casting. (A) Radiograph, (B) Photograph of outside surface.

Unfused Chaplet

Chaplets in castings are metal supports or spacers used in molds to maintain cores in their proper positions during the casting process. They are also used to support parts of the mold which are not self-supporting.

In proper casting technology, the casting metal will fuse into the chaplets. These will then generally not be readily apparent even on radiographic films, except by grinding marks on the outside surfaces of the castings.

Unfused chaplets, however, can be harmful and lead to service failures. Cracking through a cast valve body causing a steam leak as a result of an unfused chaplet is shown in Fig. 8-10.

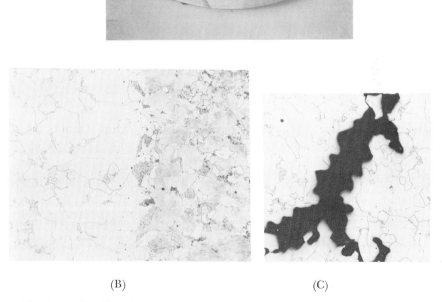

Fig. 8-11. Cracking in mild steel area embedded in $\frac{1}{2}$ Mo steel valve body (A) also showing at higher magnification (100×) decarburization (B) and cracking (C).

140 *Defects and Failures in Pressure Vessels and Piping*

METALLURGICAL NOTCHES

Dissimilar Metal Chills

To avoid or minimize shrinkage, some steel foundries have employed steel pellets or spirally wound wire in the mold to produce more uniform freezing of the molten steel. When mild steel pellets or wire imbedded in low-alloy steel castings do not melt and mix with the molten alloy steel, decarburization of the mild steel often occurs in applications involving service temperatures over 800°F. Cracking often follows. Cracking can occur even when the metallurgical composition differences are very small. Figure 8-11 illustrates decarburization and cracking in a mild steel spirally wound wire imbedded in a ½Mo steel valve body.

Fig. 8-12. Cracking in repair weld on Type 347 stainless steel casting starting from grain boundaries of stainless casting. (Magnification 5×)

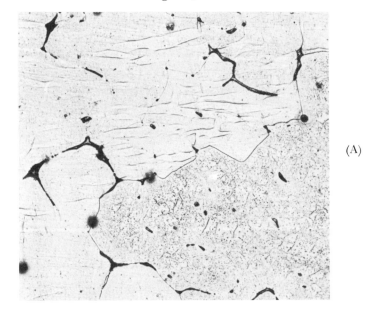

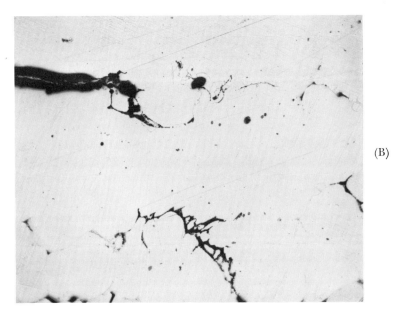

Fig. 8-13. Cracking in Type 316 (CF8M) centrifugally cast stainless steel pipe caused by inherent hot-shortness (A) which led to cracking during hot bending (B). (Magnifications (A) 100×, (B) 50×)

142 *Defects and Failures in Pressure Vessels and Piping*

Hot Shortness

Similar to wrought, plate and pipe materials, castings may be inherently hot short due to segregation along the grain boundaries. Cracking may occur during cooling, subsequent heat treatment or welding. In general, for the same chemical compositions, castings tend to show a greater susceptibility to hot shortness and hot cracking than wrought or forged products.

The effects of hot shortness in a casting are illustrated in Fig. 8-12. A weld repair was made on a Type 347 stainless steel casting in which earlier hot cracking had been removed by grinding. Cracking across the repair weld started at the grain boundaries in the casting.

In some compositions, the tendency toward hot shortness tends to increase with grain size.

Centrifugally cast pipe, particularly in stainless steel compositions, tends to exhibit a very large columnar grain structure. During hot bending or hot forming between 2100 and 1600°F, especially at the higher end of the range, cracking may be experienced in some alloys or heats susceptible to hot shortness, Fig. 8-13.

Hardness Variations

Variations in hardness discussed in the previous chapter can also be critical in castings and lead to service failures. Figure 8-14 illustrates cracking along the weld to base metal interface in a ½ Mo steel

Fig. 8-14. Cracking along weld deposit to high-hardness valve base metal interface zone (A) unetched and (B) etched to reveal the weld metal structure. (Magnification 3×)

main steam piping system. The base metal was a valve body with a somewhat higher than normal carbon content (0.26–0.31C). As is typical with castings, variations in chemical composition may intensify the susceptibility to localized hardening. As was normal procedure at the time of erection, the weld area was stress relieved at approximately 1150°F. The Brinell hardness values were found to be as follows:

Area	Brinell Hardness Values
Valve base metal	165–189
Valve heat-affected zone	231–235
Valve heat-affected zone adjacent to weld	268–290
Weld deposit	177–181

A lower carbon content in the valve base metal or a higher heat treating temperature, as might now be recommended, would have resulted in a reduced hardness differential along the weld metal to valve base metal interface.

WELD REPAIRS

The repair of casting defects by welding represents generally accepted practice. Within certain limitations, repair by welding is considered permissible by the major Codes.

Sound weld repairs can be made readily, provided that the type and magnitude of the defects have been explored and that repair welding is done under properly prepared procedures and with trained personnel.

Figure 8-15 is a radiograph of a 5Cr–½Mo alloy steel cast fitting furnished and certified to ASTM Specification A217. Although the foundry did not radiograph the casting, it made a repair by welding in the area apparent in the cross section in Fig. 8-15. Without actual radiography by the foundry, the extent of defectiveness was not explored, and excessive defects were allowed to remain. The problem was compounded by the repair being made with 25Cr–20Ni stainless steel (Type 310) electrodes. The high hardness values retained in the casting shown in Fig. 8-16 confirm that the repair welding was done without preheat and postheat treatments.

144   *Defects and Failures in Pressure Vessels and Piping*

Fig. 8-15. Photograph and print of radiograph of 5 Cr–½ Mo casting showing repair weld area.

Casting Defects

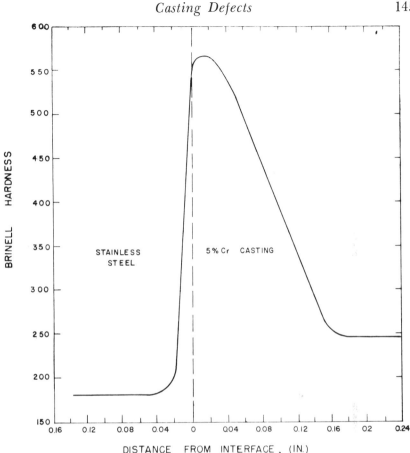

Fig. 8-16. Hardness survey confirms that repair weld was made without preheat and postheat treatments. (Hardness values were converted to Brinell from Knoop determinations.)

As described also elsewhere in this book care must be taken to select the proper welding filler metals. In some service environments, even slight differences in composition can lead to cracking and service failure. That these differences can be extremely critical is illustrated in Fig. 8-17, which shows a near failure by cracking in a repair weld. The repair weld was made by the foundry on a ½Mo steel valve (0.16C, 0.47Mo). A plain carbon steel electrode was used. The valve was then field welded with ½Mo steel electrodes into a 900°F, 900 psi main steam piping system of a steam power station. Although the

146 *Defects and Failures in Pressure Vessels and Piping*

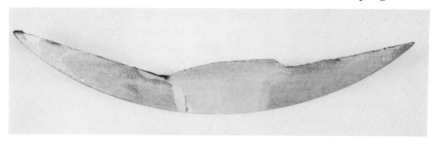

Fig. 8-17. Cracking in carbon steel repair weld in cast $\frac{1}{2}$ Mo steel valve body adjacent to $\frac{1}{2}$ Mo steel field weld joining a $\frac{1}{2}$ Mo steel pipe to the valve.

$\frac{1}{2}$ Mo dissimilarity in the repair weld might be considered inconsequential, the elevated service temperature caused carbon migration from the carbon steel repair weld into the adjacent carbon-molybdenum steel. There was sufficient loss of strength to permit initiation and propagation of a crack.

The crack was not present in the original weld. Neither radiographic examination, which originally indicated a sound weld, nor any other established test could have revealed the existing metallurgical dissimilarity, which constituted a potentially dangerous condition.

REFERENCES

1. ASTM Specification E71, "Radiographic Standards for Steel Castings," American Society for Testing and Materials, Philadelphia, Pa.
2. ASME Boiler and Pressure Vessel Code, Section I, Power Boilers, or Section VIII, Unfired Pressure Vessel, American Society of Mechanical Engineers, New York (1962).
3. Thielsch, H., "Quality Control and Service Performance," *Trans. ASME,* **86**, Series A, 451–464 (1964).
4. Polushkin, E. P., "Defects and Failures of Metals," Elsevier Publishing Co., Amsterdam, 1956.
5. Turnbull, G. K. and Wallace, J. F., "Effect of Shrinkage Porosity on Mechanical Properties of Steel Casting Sections," Steel Foundry Research Foundation, Cleveland, Ohio (Jan. 1962).

———chapter 9

NOTCH BRITTLENESS

Unexpected and sudden failures in pressure vessels, piping, bridges and other structures and failures in welded steel ships have made engineers and metallurgists aware that steels which ordinarily behave in a ductile manner may, under certain conditions, exhibit highly brittle characteristics. The majority of brittle failures have occurred in ships and bridges. A smaller number has occurred in pressure vessels and tanks. Only relatively few failures of this type have occurred in piping. In any case, the consequences of such sudden failures are extremely disturbing and can be tragic.

The importance of the problem of brittle failure is evident from the large volume of literature published on the subject, and the extensive laboratory investigations undertaken probing all aspects of steel-making practice, service environments, design factors, fabrication and welding effects, heat treatment, etc.[1-6] All of this activity had its origin in the failure of welded bridges and ships during World War II. Of over 3000 welded ships constructed at that time, 1200 sustained major hull fractures, of which 250 were considered hazardous to the safety of the vessel. At least 20 vessels have actually broken into two sections.

Investigation of the brittle failures in pressure vessels, tanks and piping generally showed that the following three factors are involved:

(1) Steel properties
(2) Design details
(3) Fabrication methods

SERVICE FAILURES

A number of examples of failures associated with notch brittleness are given in several chapters. Additional examples of serious service failures are summarized in this section.

The brittle failure at a 50 psi internal pressure of a 38½-ft diameter spherical hydrogen storage vessel in Schenectady in 1943 was attributed to (1) poor design, (2) high residual stress at the manhole welds which were cracked, (3) notch brittleness of the semi-killed steel at the 10°F temperature of failure, and (4) thermal stressing caused by a rather sudden rise of temperature of 27°F in 7 hours.[7-9]

A 40-ft diameter spherical ammonia storage vessel constructed of rimmed and semi-killed steel plates failed in Pennsylvania in 1943. The failure originated at a slight overlap of weld metal, containing also minor weld porosity. Actual failure occurred during a hammer test consisting of striking the welded seams with an 8-lb sledge hammer when the vessel was filled with water at 115 psi.[9]

In 1944, in West Virginia, a spherical pressure vessel normally containing liquefied gas failed in a brittle manner during a hydrostatic test at 98 psi. Investigation indicated that the failure started in the supporting column at a point below where it was attached to the shell.[9]

In 1944, the failure of a cylindrical liquefied gas-pressure vessel in Cleveland, Ohio, resulted in the death of 128 people. The service temperature of the pressure vessel was −260°F. The plate material was a 3½% nickel low-carbon steel welded with Type 310 (25Cr–20-Ni) stainless steel electrodes. Although the origin of the failure was not definitely established, the plate material had been furnished in the as-rolled condition, and had not been normalized as had been specified.[1,9,11]

In January 1945, a 43-ft high by 3-ft 7-in. ID methane column constructed of firebox quality carbon steel failed in a brittle manner after 15 years of service. The normal operating temperature was −166°F at the top of the column, and −94°F at the bottom. The gage pressure was 125 psi. Incomplete fusion and lack of penetration in a weld near the liquid line near the bottom of the column and stress concentrations produced by side connecting openings were believed to have contributed substantially to the failure. However,

Notch Brittleness 149

the extreme notch brittleness of the steel at the operating temperature was considered the primary criterion. Izod impact test values of the steel varied from 39 to 59 ft-lb at room temperature down to 1 to 3 ft-lb at the operating temperatures.[1]

In 1947, a crude oil storage tank failed in a brittle manner in the midwest. The failure occurred while the tank was being filled with crude oil. The oil temperature was 43°F and the air temperature was 0°F. The failure originated at the square upper corner of the clean-out door, representing a design notch similar to notch conditions which caused failures of Liberty Ships starting at hatch corners. The steels involved were rimmed grades and notch sensitive. Weld deposits around the corners were also found to be of poor quality.[1]

Surface irregularities, undercutting and generally poor welding appear to have caused or contributed to the failure of a 10,000 cu m. oil storage tank in Normandy, France, during the 1950-1951 winter.[12]

A 177-ft high by 12-ft diameter bricklined steel stack of welded construction failed in November 1951 in Chicago after 10 years of service. The temperature had dropped to about 12°F during the two weeks preceding discovery of the crack. Fairly high winds at velocities of up to 24 mph and higher caused some vibrations in the stack. Investigation showed that (1) the steel behaved in a brittle manner at the temperatures preceding the failure, (2) an increase in the hoop stress was caused in the steel shell by temperature changes and expansion of the lining, and (3) additional stresses were produced by the wind and vibrations. Other stacks in the same area of steels of lower transition temperatures did not fail under the same service conditions.[1, 13]

In 1952, two large 140-ft and 150-ft diameter storage tanks failed in Fawley, England. Failure in the 140-ft crude oil storage tank started at a faulty patch weld where an inspection probe had been removed on the first horizontal weld seam. The failure in the 150-ft crude oil storage tank started at an improperly repaired weld crack. The crack was old, as its surfaces were coated with a black oxide film from subsequent welding operations. Poor welding, therefore, was considered to be the primary cause of these tank failures.[1, 14]

In 1952, three empty oil tanks of floating roof design 144 ft in diameter by 45 ft high, failed in Europe. The seam welds had been chipped flush and considerable hammering had been done to correct distortion. Several weeks after erection, when the temperature fell to 25°F, cracks developed in the chipped and hammered surface area, and extended transversely across the welds. Failure was considered to be caused by (1) the local notches caused by chipping and hammering, (2) the residual tensile welding stresses, and (3) the increased notch sensitivity of the steel caused by the fall in temperature and the work hardening produced by the hammering.[15]

NOTCH BRITTLENESS

The common denominator in all the failures described was notch brittleness, generally occurring at particularly low temperatures.

The majority of failures described in the other chapters are generally associated with a ductile state where the fracture is produced at loads equivalent to a general stress equal to the ultimate tensile strength of the material. In most instances, crack growth has been gradual, producing initially only a small leak.

In notch brittle materials, fractures can be initiated at notches or weld defects at stresses less than yield point load. Once initiated, such fractures will continue to extend as long as a tensile stress of sufficient magnitude is present. Crack propagation in steels is possible at stress levels as low as 8000 psi.[16]

It is not unusual for engineers to ascribe the brittle fractures associated with these failures to such metallurgical causes as poor steel quality or an impairment of quality from welding operations. However, actual examination of most sudden failures of steel fabrication has revealed that superimposed mechanical factors were intimately associated with these brittle failures. In general, brittle failures have been found to be associated with inferior design, faulty workmanship including careless or improper welding practice, use of a steel which is notch-sensitive under the particular operating (or testing) conditions, or a combination of these influences. Whereas the origin of many brittle failures can be traced to weld defects, the crack or cracks often propagate in the plate or pipe material, and not in the weld. Particularly in piping which is welded from the outside only, severe notches may result from lack of penetration or cracks in the root of the weld. Even thorough radiographic inspection may not reveal lack of penetration or cracking in the root of the weld along the pipe inside surface.

DUCTILE VS BRITTLE BEHAVIOR

Characteristics

Steel is generally considered to be a ductile material. When overloaded, it usually gives warning by flowing plastically, i.e., bulging, stretching, bending or necking, before rupturing. Contrary to expectation, however, steels sometimes rupture without prior evidence of distress. Such brittle failures are accompanied by but little plastic deformation, and the energy required to propagate the fracture appears to be quite low. Under certain conditions, steel may shatter like glass. In piping, this extreme behavior generally occurs only at low temperatures.

Three conditions control this tendency for steel to behave in a brittle fashion. These include (1) high stress concentrations, i.e., notches, nicks, scratches, internal flaws or sharp changes in geometry; (2) a high rate of straining; and (3) a low environmental temperature. These three factors are so interrelated that the determination of the effect of any one of them provides an indication of how the steel will react to the intensification of either or both of the others. The effect of lowering the testing temperature is the condition most convenient to measure quantitatively. Consequently, the transition from ductile to brittle behavior of a steel is generally expressed in terms of temperature.

The transition temperature for any steel is the temperature above which the steel behaves in a predominantly ductile manner and below which it behaves in a predominantly brittle manner. Steel with a high transition temperature is more likely to behave in a brittle manner during fabrication or in service.

Figure 9-1 shows the fracture surfaces of a $10\frac{3}{4}$-in. OD by 1.250-in. wall ASTM A106 Grade B pipe material which broke in a brittle manner during shop bending of the pipe at room temperature (at about 75°). The material, in larger as well as in smaller diameters, normally can be cold bent readily. Charpy impact tests showed that the pipe sections which failed had a significantly higher transition

152 *Defects and Failures in Pressure Vessels and Piping*

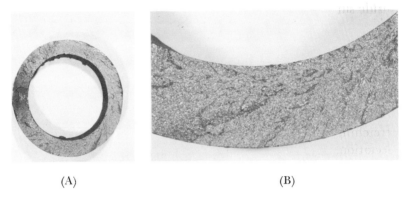

(A)   (B)

Fig. 9-1. Brittle failure in 10¾-in. OD by 1.250-in. wall carbon steel pipe during cold bending. (A) Reduced to ⅕ size.

temperature than is normal for ASTM A106 Grade B pipe material, Fig. 9-2.

It follows that a steel with a low transition temperature is more likely to behave in a ductile manner and, therefore, steels with low transition temperatures are generally preferred for service involving severe stress concentrations, impact loading, low temperatures or combinations of the three.

Metallurgical factors, such as deoxidation practice, chemical composition, rolling, forging or extruding practice, and subsequent heat treatment, influence the transition temperature of steel. In carbon steel materials, under the worst conditions, the transition temperature may be above 200°F or under the best condtions, below −100°F.

Steels treated in accordance with most favorable deoxidation practice are those which are fully killed. In plate and pipe steels, deoxidation is generally accomplished with sufficient silicon to provide about 0.10 to 0.20% of silicon in the steel. Aluminum has been used also as a deoxidizer, although less in recent years, than several decades ago. The amount of aluminum normally used causes the retention of only a few hundredths of a per cent residual aluminum in the steel. Such steels are sometimes referred to as being made in accordance with fine-grain melting practice.

Carbon influences the transition temperature of as-rolled or normalized steels unfavorably. Its upper limit in plain carbon steels is generally accepted as about 0.25%, and in low-alloy steels as about

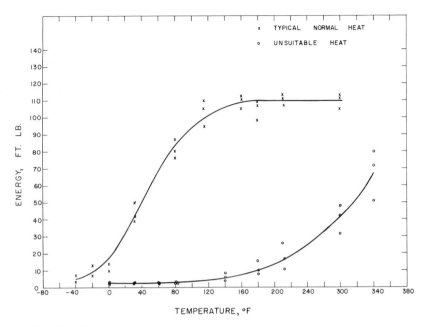

Fig. 9-2. Charpy V-notch transition-temperature curves of ASTM A106 Grade B pipe considered suitable and unsuitable for cold bending (0.24 C, 0.62 Mn, 0.22 Si, 0.011 P, 0.027 S).

0.20% or even lower. High ratios of manganese to carbon may be beneficial. Most elements other than those used for deoxidation raise the transition temperature with the notable exception of nickel, which lowers appreciably the transition temperature of carbon steel. Austenitic chromium-nickel stainless steel and some high-nickel steels do not show transition at temperatures even as low as −325°F to −425°F.

Steels which have been fully annealed are in the poorest condition to resist embrittlement. Normalizing provides improvement. Frequently, further benefit is derived from tempering after normalizing, or from stress relieving after welding. Optimum properties are obtained by fully quenching and tempering to moderate strength levels. Such treatment is seldom applied to pipe steels and has not received recognition in the ASA Code for Pressure Piping and the ASME Boiler and Pressure Vessel Code.

The presence of notches or other stress-concentration factors is of very considerable importance. Sharp notches in welded joints may encourage failure at loads well below those permitted by design. Any aggravated notch, just like a severe geometric shape change, can result in a crack at relatively low loads.

Considerable quantities of plate and pipe steels of rimmed or semi-killed types (for example, ASTM A53) have operated satisfactorily in service for long periods although they have a higher transition temperature than fully killed steels (for example, ASTM A106). These steels offer cost advantages and are usually utilized in shock-free service when failure is unlikely. However, even in the absence of shock, pressure vessels and piping operating under high stress should be made of steels which are tough at the lowest service temperature, such as those commercially available under ASTM Specification A106. This is particularly important for pressure vessels, tanks and piping located where failure would endanger life and property.

Brittle failures rarely occur in piping. Nevertheless, their prevention is of paramount importance because economics and fabrication limitations rarely favor construction of a perfect piping system. Pri-

Table 9-1. Low-temperature Limitations for Various Plate and Piping Materials.

Low Temp. Limit, °F	Material and Suitable ASTM Designation	Comments
Zero	Mild steel (A53, A120, A135)	No requirements other than suitable pressure rating
−20	Mild steel (A53, A106, A135)	Reduce pressure rating 1% for each F deg. below zero, or Charpy impact test, 15 ft-lb at design temperature
−50	Killed or limited carbon steel (A201, A212, A333, Gr. 0)	Charpy impact test, 15 ft-lb at design temperature
−150	$3\frac{1}{2}$% Ni-steel (A203, Gr. D, E; A333, Gr. 3; A410)	Charpy impact test, 15 ft-lb at design temperature
−320	9% Ni Steel (A353)	Charpy impact test, 15 ft-lb at design temperature
−420	Austenitic stainless steel (Types 304, 316, etc.)	Austenitic stainless steels are not normally susceptible to notch-brittleness
No Limit	Copper, brass, aluminum	Not normally susceptible to notch-brittleness

mary emphasis should be placed upon the selection of steels of suitable quality for their intended use, and followed by design and fabrication practices that will hold stress concentrations to an acceptable minimum.

Low-temperature Service

Table 9-1 indicates the low temperature limitations normally considered applicable to various plate and piping materials. Low-alloy steels may be used below zero when they have Charpy Keyhole impact values of at least 15 ft-lb at the lowest design temperature. Austenitic stainless steels with a limited carbon content, copper and copper alloys, and aluminum do not experience transitions in impact strength from ductile to brittle fracture and, therefore, may be used for low temperatures without pressure rating penalties.

Low temperature piping is generally covered with thermal insulation which helps provide protection from external impact blows or shock. This, however, is not sufficient insurance against the type of damage that could result if a pipe should fracture.

Nature of Brittle Behavior

Brittle behavior in steel may result from a number of factors including (1) rapid rate of straining, (2) the presence of multidirectional stresses, or notches causing restraint such as surface defects, discontinuities, incomplete welds, underbead cracks, microcracks, sharp re-entrant corners, etc., (3) low operating temperature, or (4) a combination of these factors.

The brittle behavior of a steel is generally evidenced by sudden fracture and sometimes by shattering. The initial cracking usually will propagate very rapidly and, under certain conditions, the rate of propagation is practically infinite. Relatively little energy is required for propagation. At failure, the surface of the fracture appears bright, granular and crystalline with the cross section showing little or no evidence of necking or plastic deformation. A typical example is the notched specimen shown in Fig. 9-3. Such fractures are generally called *cleavage* fractures.

In contrast, a steel which behaves in a ductile manner generally will fail gradually, i.e., at a much slower rate. A high energy impact is required for crack propagation. Moreover, ductile failures in pres-

156 *Defects and Failures in Pressure Vessels and Piping*

Fig. 9-3. Brittle cleavage fracture in mild steel—bright, granular and crystalline. Plastic deformation is absent.

sure vessels, pipe or tubes usually are preceded by local bulging, i.e., by plastic deformation, or flow in the material because of shearing forces. Such *shear* fractures, as in the notched specimen shown in Fig. 9-4, will appear relatively dull and fibrous and show a measurable and often considerable contraction of area across the fracture section.

Evaluation of Brittle Behavior

The conventional static tension and bend tests do not adequately differentiate between steels of varying susceptibilities to brittle behavior. This has created the desire for a simple test that could be used to select steels that would be considered "safe" for use under any predictable conditions. Such a test should permit selection of a particular steel for a particular type of element in the structure for use under normal and, possibly, for an occasionally unusual service condition or environment which the structure may encounter.

Unfortunately, there is no single test that fulfills these requirements. This is because the common tests measure the mechanical properties of the steel under the particular conditions imposed by the test method, and not its behavior in an actual structure as influenced by such factors as design, workmanship, surface notches, welding quality, restraints and stress distribution. Nevertheless, there are several accepted tests that are extremely useful because they permit comparison among the various types and grades of steel. Moreover, these tests, which usually require notched-bar or notched-

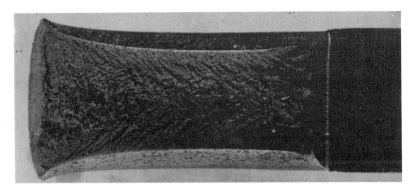

Fig. 9-4. Ductile shear fracture in mild steel—dull and fibrous. Plastic deformation is present.

plate specimens, provide some knowledge of brittle behavior and the approximate relation of such factors as thickness of the steel, chemical composition, deoxidation practice, sharpness of notches, heat treatment, etc., to fracture characteristics.

The determination of the temperature at which a steel may become susceptible to brittle failure under certain conditions is based on one of three testing categories: (1) impact energy, (2) fracture appearance, and (3) notch ductility. These criteria are also referred to as energy transition, fracture transition, and ductility transition. For Charpy Keyhole and V-notch tests, these relationships are illustrated in Fig. 9-5.

In the ductility transition criterion, the ductility of the steel at the notch apex is of primary concern. In the fracture transition, the criterion depends upon crack propagation. Thus, the ductility transition is concerned primarily with fracture initiation. It indicates the temperature above which a brittle crack will initiate only after appreciable plastic flow at the base notch and below which a brittle crack will initiate without significant evidence of notch ductility. The fracture transition indicates the change from a fibrous shear-type to a cleavage-type fracture. Cleavage-type fractures do not take place above the fracture transition temperature. Thus, the fracture appearance transition temperature is normally somewhat higher than the ductility transition temperature.

The transition temperature of any one grade or type of steel as

158 *Defects and Failures in Pressure Vessels and Piping*

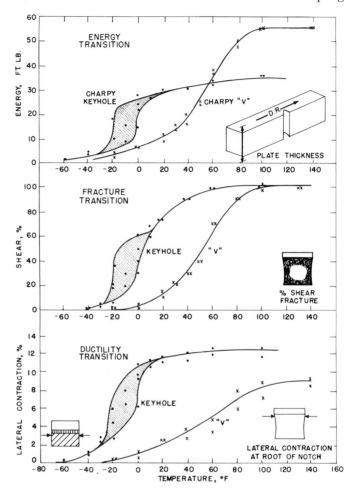

Fig. 9-5. Relationship between various transition-temperature test criteria. (Pellini)

determined by one type of test generally should not be compared to that of another grade of steel unless the same criterion of brittleness has been used. Under the same type of test, the transition temperatures obtained are of only qualitative value in assuring that the steel which exhibits the lower transition temperature is less likely to fail in a brittle manner than the steel showing a higher transition temperature. For any particular service, however, other factors being equal, the steel with the lower transition temperature should be

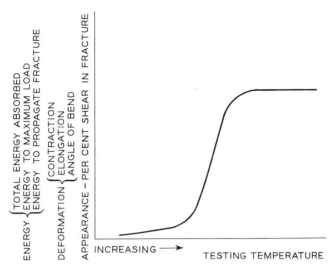

Fig. 9-6. Schematic illustration of transition-temperature range.

preferred because it offers additional safety, even though the limiting low service temperature of the structure may not be precisely known.

Effects of Temperature

The transition from ductile to brittle behavior of carbon and low-alloy steels is not a phenomenon only occasionally noted; it is an inherent property which can be demonstrated by a number of accepted test methods. For example, Fig. 9-6 illustrates schematically this transition as brought about by a decrease in the testing temperature. It should be noted that since several criteria may be employed to evaluate the transition temperature, its position with respect to the temperature scale may be expected to vary with the criterion selected.

Transition Temperature

Various tests have been developed to determine whether a steel under certain test conditions will fail definitely in a ductile or brittle manner. Each of these conditions actually shows a scatter region or a transition-temperature range, Fig. 9-7, within which a steel under

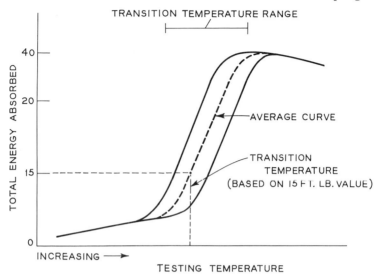

Fig. 9-7. Transition-temperature range and transition temperature in Charpy impact test.

the same set of test conditions will exhibit either ductile or brittle behavior, or both.

Usually, some empirically established point within the transition-temperature range is selected and is thereafter designated as the transition temperature for the steel tested. This temperature can be roughly defined as the temperature above which the steel under the particular test conditions will behave in a predominantly ductile manner and below which the steel will exhibit a predominantly brittle behavior.

For example, when the appearance of the fracture is the criterion, the transition temperature may be arbitrarily designated as the temperature at which half of the fracture surface shows a cleavage failure while the other half exhibits a shear failure. When impact energy absorption is the criterion which represents the most widely accepted method, some arbitrary energy value is generally used. For example, in the Charpy test, the temperature corresponding to the 15 ft-lb value is often designated arbitrarily as the transition temperature. It must be noted that at this value, the fracture may be predominantly of the cleavage type. This last method is illustrated in Fig. 9-7.

Notch Brittleness

Other criteria are the temperature at which the initial appearance of cleavage (brittle) fracture is observed and the temperature at which the energy absorbed is half of the maximum energy value obtained over the testing range.

It should be recognized that there exists no one "transition temperature" for a given steel except for a particular set of conditions and one criterion of brittleness. Each particular forged or extruded pipe or plate thickness rolled from a given steel will exhibit a range of temperatures in which the fracture shifts from tough to brittle. This range may be varied considerably by changing the metallurgical and mechanical properties of a steel.

Tests conducted on gas line pipe [17] employing Charpy V-notch test specimens indicated an apparent correlation between fracture speed, the number of fractures traveling simultaneously, and the position on the Charpy V-notch curve. As the pipe rupture test temperatures were dropped from the top to the bottom of the Charpy test curve, the number of fractures and the crack propagation speed increased. The test results are illustrated in Fig. 9-8. The Charpy V-notch curve represents the results of four sets of tests and shows also the temperatures of four pipe rupture tests. The fracture in Test 2 was completely shear and traveled at the lowest speed (650 ft/sec). Fractures in the other three tests were all cleavage in nature, and traveled at increasing velocities as the temperature became lower.

The relation of shear area to Charpy V-notch impact energy may not always be the same as illustrated in Fig. 9-8. For example, in a 60,000 psi average yield strength pipe steel for gas-transmission service, the 50% shear fracture area in the Charpy V-notch test correlated with the observed arrest of fracture in an initial service failure at 125 to 150°F. Pipe lengths capable of absorbing more than 30 ft-lb were on the average observed to fail in a partly ductile manner, and did not produce fragments. Pipe lengths capable of absorbing less than 30 ft-lb failed in a completely cleavage manner producing one or more fragments.[18]

Effects of Notches and Notch Sensitivity

The sudden brittle failures which occur without measurable deformation are generally ascribed to the notch sensitivity of the steel at the operating temperatures to which the steel was exposed. Al-

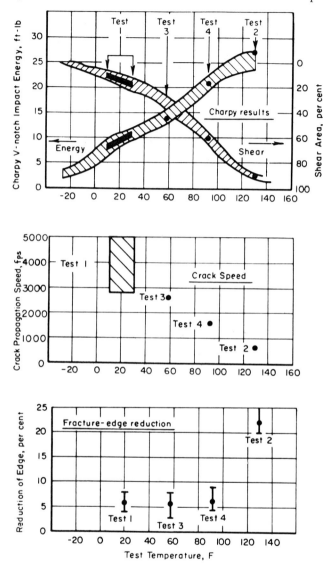

Fig. 9-8. Summary of results of experiments on gas-transmission pipe involving correlations between rupture tests and Charpy V-notch tests.[17]

though not always obvious, actual notches or notch effects are present in all structures. They may consist of minute surface or subsurface cracks, surface laps or scabs, visible scratches, abrupt shape changes such as sharp corners, tool marks, grooves from drawing dies, edges, etc., or fabrication defects, such as arc strikes, etc.

Notch sensitivity is usually associated with an inability of the steel to deform in a plastic manner (that is to flow) underneath the notch. This resistance to flow is increased by the triaxial state of stress induced under a notch by a tensile stress. The condition is accentuated as the thickness of the steel increases. Notches also are stress raisers. The greater the sharpness of the notch, the greater will be the degree of restraint, the more severe the degree of restraint, the more severe will be the stresses as to both triaxiality and magnitude, and the higher will be the transition temperature. The effects of notches of varying severity on the brittle behavior of Charpy test specimens are illustrated in Fig. 9-9. This is the basis of the general concept that the transition-temperature range or the arbitrarily selected transition temperature will increase with the severity of the notch. In other words, a steel which contains extremely severe notches will fail in

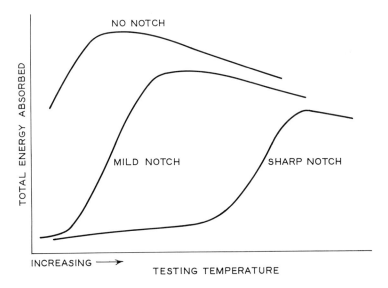

Fig. 9-9. Effects of notch severity on the transition temperature.

a brittle manner at higher ambient temperatures than if less severe notches were present.

It should also be noted that in instances where service conditions impose a triaxial state of stress, as is true of internally loaded vessels and piping, brittle behavior may be encountered even in the absence of notches.

Fracture Analysis Diagram

The understanding of fracture initiation and crack arrest is facilitated by fracture analysis diagrams.[19-22] These diagrams are related to the service temperature by the determination of one of several significant reference transition temperatures, Fig. 9-10.[19, 20, 22] These temperatures may be determined by the drop-weight test, or by the Charpy V-notch test, as correlated to the drop weight test. The reference temperature is designated as the *nil ductility transition* (NDT).

In a flaw-free steel, the tensile and yield strengths would increase gradually with decreasing temperature. Since the increase in yield strength is greater than the increase in tensile strength, at some very

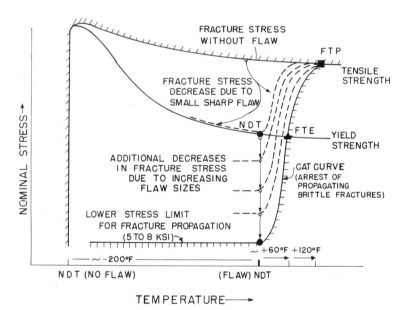

Fig. 9-10. Transition-temperature characteristics of steels.[19, 22]

low temperatures, the yield and tensile strength values would coincide. At this temperature, the ductility becomes essentially zero. Thus, for a steel free of flaws, this temperature would be the NDT temperature.[19, 22] If a small sharp flaw is notched into the test specimen, a decrease in fracture strength results, as indicated by the dotted curve. The continuation of the dotted curve to lower temperatures is represented by the yield strength curve. The temperature at which the decreasing fracture initiation (small flaw) becomes continuous with the yield strength curve of the steel is defined as the NDT. The arrow pointing down from the NDT point indicates that increasing the flaw size results in fracture at lower nominal stresses.

A second important curve approximates the fracture *arrest* relationship between stress and temperature. This *crack arrest curve* represents the temperature of arrest of a propagating brittle fracture for various levels of applied nominal stress. The crack arrest temperature (CAT) for yield stress loading is then defined as the *fracture transition elastic* (FTE). It is the highest temperature of fracture propagation for purely elastic loads. The *fracture transition plastic* (FTP) represents the temperature above which fractures occur en-

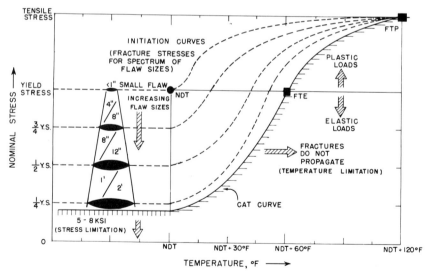

Fig. 9-11. Generalized fracture-analysis diagram, as referenced by the NDT temperature.[19, 22]

tirely by shear; i.e., they show no center regions of cleavage fracture. The stress required for fracture approximates the tensile strength of the steel.

The lower horizontal level of the curve represents the stress level, normally between 5000 and 8000 psi, below which fracture propagation does not occur. This is because the minimum elastic strain energy release required for continued propagation of brittle fractures is not attained.[19]

Figure 9-11 illustrates the effect of flaw sizes on fracture initiation at various levels of nominal stress. This diagram shows that for a given level of stress, larger flaw sizes will be required for fracture initiation above the NDT temperature. For example, at 75% yield strength loads, a flaw of the order of 8 to 10 in. may be sufficient to initiate fracture below the NDT temperature. However, at a temperature of NDT + 20 or 30°F, a flaw of 1½ or 2 times this size may be required for initiation. To the right of the CAT curve, there is "no flaw" that is of sufficient size to initiate fracture because propagation is not possible.[19]

The correlation of test results with actual failures involving leaks or catastrophic failures has been quite good.[20, 23] Catastrophic failures

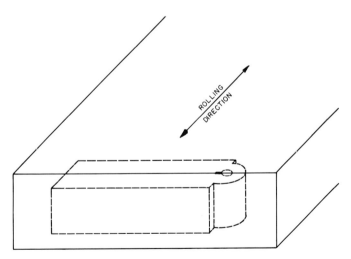

Fig. 9-12. Sketch illustrating specimen used in isothermal crack-arrest test to obtain the results shown in Fig. 9-13.[27] The test specimen is machined from center of plate.

Notch Brittleness

with conditions which appear to the right of the CAT curve can only occur where an exceptionally long or deep flaw reduces the section sufficiently to cause instability in the unfractured region.[23]

Isothermal Crack Arrest Concept

In England, the *isothermal crack arrest temperature* concept has been used in the acceptability of steel materials. It is based on the assumption that in large structures, it is impossible to avoid notches such as cracks and other flaws. Consequently, crack initiation is considered potentially possible. Reliance is placed on service conditions insuring the nonpropagation of the inherent cracks or defects.[24-26]

The isothermal crack arrest temperature is the lowest temperature at which the crack fails to sever the test specimen. In this test, a brittle crack is artifically initiated at one end of a notched specimen, (Fig. 9-12) held under known stress and temperature conditions. A temperature gradient is maintained across the specimen. The crack propagates across the specimen and stops at the crack arrest temperature (CAT).

By plotting the nil ductility transition (NDT) against the isothermal crack arrest temperature, the relation shown in Fig. 9-13 is obtained.[27]

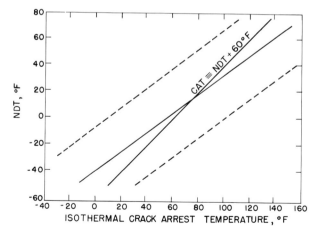

Fig. 9-13. Relation of nil ductility transition (NDT) to isothermal crack-arrest temperature and comparison crack-arrest temperatures. (NDT + 60°F) obtained from drop weight tests.[27]

The results of this test generally appear to provide fairly good correlation to the drop weight test results plotted in the fracture analysis diagram. To illustrate the correlation, the NDT + 60°F curve is included in Fig. 9-13.[27]

TEST METHODS

Small Scale Test Specimens

Various "small-scale" tests have been developed to evaluate the notch sensitivity of steel. These tests may be divided roughly into three categories consisting of single-blow notched-bar impact tests, slow-bend notch tests and notched tensile tests. The latter include the symmetrically loaded type of specimens such as the edge-notched plate test and the asymmetrically loaded type of specimens such as the Navy Tear Test.

"Small-scale" notched-bar impact tests are by far the most widely used to evaluate steel plate and piping materials. They are generally made for one of three primary purposes: (1) to examine and correlate the metallurgical characteristics of various types of steels, (2) to inspect the steel (approve or reject) in accordance with specification and code requirements, and (3) to indicate in so far as possible the service behavior of a particular steel in a finished structure.

Of the various notch-impact specimens discussed in the literature, the Charpy keyhole and V-notch specimens are most widely used, and are specified in various materials and construction specifications. Whereas ten years ago the Charpy keyhole specimens were more widely used, the V-notch type is increasingly preferred and used in the reference literature.[28] The advantage of the V-notch specimen is that it develops a smooth transition temperature curve, thus providing a useful correlation of an ft-lb energy index point to a reference transition temperature. On the other hand, the Charpy Keyhole energy curve is discontinuous, involving high energy and low energy branches connected by a transition scatter band. Such scatter data make correlation difficult.[29]

The fracture analysis diagram is based on *drop weight impact tests*[30] now recognized as a standard test.[31] The test specimen has applied to its surface a *crack-starter weld*. This represents a 2½ in. long by ½ in. wide centrally located weld bead. A notch is cut into

Notch Brittleness

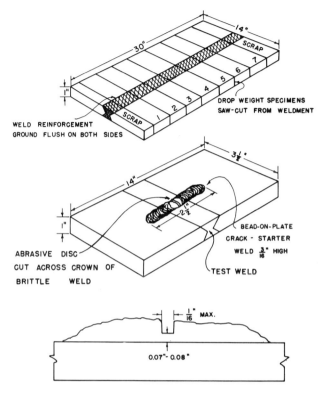

Fig. 9-14. Weld and notch details for drop weight test specimen.

the bead, Fig. 9-14. After cooling the specimen to the test temperatures, a weight is dropped onto the specimen, Fig. 9-15. Generally, six to eight specimens are required to establish a significant NDT temperature. Depending on the results of the first test, subsequent tests are conducted in steps of 40°F until a "break" and a "no-break" performance is obtained. A test is then conducted at the 20°F temperature midway between the "break" and "no-break" temperatures. This 20°F temperature range is then halved again at 10°F to determine the lowest "no-break" temperature.

A number of notched tension test specimens have also been used to provide correlation to service failures. Among various tests suggested, two of the more widely used specimens are illustrated in Fig. 9-16. In the Robertson test,[16] a crack is started by impact in a knob on the edge of a plate specimen in tension. The knob is cooled with

170 *Defects and Failures in Pressure Vessels and Piping*

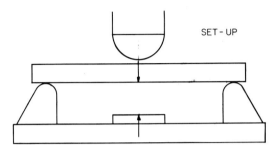

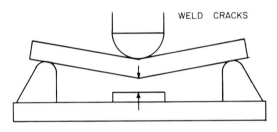

YIELD POINT LOADING IN PRESENCE OF SMALL
CRACK IS TERMINATED BY CONTACT WITH STOP

Fig. 9-15. Sketch illustrating essential features
of drop-weight test method.[30, 31]

liquid air to minimize the impact energy required to start the crack. By heating the opposite edge of the plate, a temperature gradient is provided. At various stress levels, the temperature in the plate is determined at which crack propagation stops. The temperature represents the crack arrest temperature CAT. A slight modification of the specimen shown in Fig. 9-12 is used to determine the isothermal crack arrest temperature shown in Fig. 9-13.

In the Esso test,[32,33] Fig. 9-16, a wedge is driven into a previously prepared transverse notch. When the specimen at a particular and uniform temperature is loaded to a sufficient stress, the crack will propagate across the entire width of the specimen. If the stress is too low, or for a specific stress the temperature is too high, the crack will not propagate. By varying the stress at a constant temperature, or the temperature at a constant stress, a stress-temperature relation is determined indicating when cracks will either propagate or will not prop-

Notch Brittleness

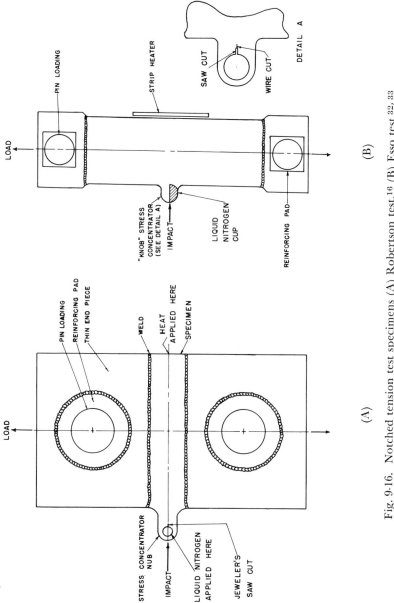

Fig. 9-16. Notched tension test specimens (A) Robertson test,[16] (B) Esso test.[32, 33]

agate. The impact on the wedge is supplied by a rivet shot from a gun using a 45-caliber cartridge with variable powder charge.

Numerous other tests have been discussed in the literature. Many publications also illustrate correlations of other special tests to standard notched-bar impact tests, such as Charpy V-notch tests. In some instances, the test results are in good agreement, in others, the correlations are poor. Variables such as steel-making practice, thermal history, etc., also influence the success of the correlations obtained.

The "acceptance" tests made in accordance with requirements of design codes and materials specifications, and the tests made to obtain an indication of service behavior are of major interest to the engineer. They are, of course, also useful to the metallurgist, as they guide him in his efforts to obtain steel of the desired properties.

Determination of the effects of metallurgical variables on the test properties of steels is relatively simple, but to interpret the effect of these variables in terms of probable service behavior is extremely difficult. One reason for this difficulty is that, in many respects, the steel in a relatively rigid pressure vessel or piping system does not behave in the same manner as it does in small-scale laboratory test specimens. Moreover, in vessels, tanks and piping systems, certain mechanical shape and assembly factors and adverse effects of fabrication procedures and workmanship may come into play. Consequently, the results obtained with the conventional notched-bar testing procedures on small specimens are not necessarily the same that would be obtained from tests of full-scale structures of the same steel. Nevertheless, some large-scale tests such as direct explosion tests have been suggested for evaluating the effects of welding and other fabricating processes. The initial results of direct explosion tests indicate significant differences in the performance of steels which have been welded by different procedures. Ultimately, these tests may prove to be useful for determining the proper welding procedure to obtain low-transition temperatures in vessel and pipe weldments.

Code-Test Requirements

For carbon and low-alloy steels for use below $-20°F$, the 1962 edition of the Unfired Pressure Vessel Code, Section VIII of the ASME Boiler and Pressure Vessel Code, requires that three Charpy Keyhole specimens exhibit an average impact value of 15 ft-lb at the lowest

Notch Brittleness

operating temperature, with only one of the specimens permitted to show a minimum of 10 ft-lb. The V-notch Charpy and the Izod specimens are not recognized by the 1962 Unfired Pressure Vessel Code. Similarly, ASTM Specification A300-58, A333-63T, A334-63T, A350-61T and A420-63T for carbon and alloy steel pipe materials for service at low temperatures requires impact testing with Charpy specimens having either a Keyhole or equivalent slot notch. Tests must be made at the minimum temperatures shown in Table 9-2.

TABLE 9-2. MINIMUM IMPACT TESTING TEMPERATURES FOR VARIOUS LOW-TEMPERATURE STEELS.

Material	Plate Grades	Pipe Grades	Temperature Min. °F
Carbon Steel	A201, A212	0	−50
2¼Ni Steel	A203, A & B		−75
3½Ni Steel	A203, D & E	3	−150
Cr-Cu-Ni Steel	A410	4	−150
4½Ni Steel		5	−150
9Ni Steel	A353		−320

These specifications also recognize the effects of specimen size, as shown in Table 9-3. The average of three impact tests specimens must meet the values given in Table 9-3.

TABLE 9-3. IMPACT REQUIREMENTS IN ASTM SPECIFICATIONS A300, A333, A334, A350, A420.

Size of Specimen, mm	Minimum Average Notched Bar Impact Value of Each Set of Three Specimens, ft-lb	Minimum Notched Bar Impact Value of One Specimen Only of a Set, ft-lb
10 by 10	15	10
10 by 7.5	12.5	8.5
10 by 5	10	7.0
10 by 2.5	5	3.5

Significance of Test Data

Mechanical-property requirements of standard specifications for steel plate and pipe materials supply only limited information to the designer, although such requirements are useful in purchasing and classifying materials on the basis of tensile and flattening properties. One reason tensile test data are of limited application in design is that the tension test is uniaxial, whereas the hazard of brittle failure is due to the existence of multiaxial stresses and stress concentrations. The static tensile strength, yield strength and ductility of the unnotched tensile bars at ambient temperatures and the flattening characteristics represent a relative quality factor rather than a value that applies to design. This is substantiated by the fact that certain special alloy steels, which in the conventional room-temperature tensile test have elongation values of slightly under 5% have given satisfactory service in many applications. However, these special steels had been processed to develop transition temperatures considerably lower than that of mild steel having conventional tensile ductility values above 25%. In the applications involved, steels of higher transition temperatures would not have performed satisfactorily no matter how high their tensile test ductility at room temperature.

It is not intended to discount the need for ductility in steel for satisfactory performance in pressure vessels and piping systems. However, ductility at room temperature as determined in the static tensile test does not necessarily indicate that the steel will behave in a ductile manner under different conditions of stressing. For instance, large cross-sectional areas and undesirable design features may cause mechanical restraint. As mechanical restraint opposes plastic deformation, the steel will fail in a brittle manner if restraint is sufficiently severe and if the temperature is sufficiently low. The major difficulty is that there is no accepted method of correlating ductility and toughness with other mechanical properties obtained from standard tensile or notched-bar tests that can be applied to assure the most efficient use of a steel in a pressure vessel or piping system.

Ductility and Fracture Transition

Test results obtained with many types of specimens usually will show two transition temperatures instead of one. This is shown

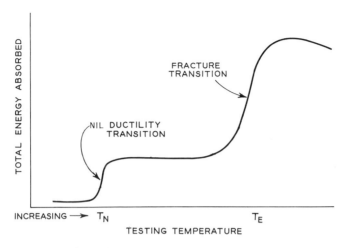

Fig. 9-17. Ductile and fracture transitions.

schematically by the solid curve in Fig. 9-17. The two transition temperatures are identified as fracture transition and nil-ductility transition. The drop in energy at T_E (the fracture transition) is generally associated with the change in fracture appearance over the fractured surface which occurs in this temperature region. The drop at T_N (the nil-ductility transition) is accompanied often by a change in fracture appearance of a small zone at the base of the notch.

The temperature at which the nil-ductility transition occurs is very much dependent upon notch geometry and the influence of plastic straining prior to failure. As the notch is made sharper and deeper, the strain is more localized and the degree of triaxiality of the state of stress is greater. This leads to higher nil-ductility transition temperatures.

The fracture-transition temperature range, on the other hand, is less sensitive to notch geometry as specimens with different size notches may show widely different ductility transitions without exhibiting similar differences in the fracture-transition temperatures. As an example, in Fig. 9-18, are shown schematically the transition curves for (A) large, centrally notched, wide-plate specimens and for (B) small notched specimens.

Both transition-temperature ranges have a bearing on the behavior of the steel. Whereas the nil-ductility transition (T_N) is associated

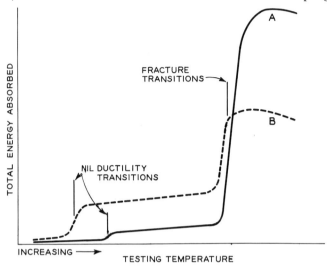

Fig. 9-18. Transition temperature of large, centrally notched wide plate specimens (Curve A) and of small-notched specimens (Curve B).

with the initiation of a crack, the fracture transition (T_E) is associated with its propagation.

FACTORS DETERMINING TRANSITION TEMPERATURES

The factors which influence the transition-temperature range of a steel may be separated into metallurgical factors and mechanical factors. The metallurgical factors are characteristic of the steel itself, whereas the mechanical factors depend upon the service or testing environments as well as many effects produced by fabricating procedures. An analysis and understanding of the various factors is often highly important, particularly when it is desirable to make the most efficient use of a particular steel.

Metallurgical Factors

Composition. Carbon and nitrogen are considered to be among the most important elements which raise the transition temperature of steel in the as-rolled or normalized condition. Oxygen and phosphorus in quantities greater than normally tolerated, and silicon in percentages greater than required for deoxidation, also may raise the transition temperature. In fact, most additions usually made to steel raise the transition temperature. The additions that will lower the transition temperature are nickel and under certain conditions, manganese. Nickel is frequently added to low-carbon steel when a transition temperature below that obtainable with carbon steel is desired. Aluminum, in the range of quantities used for deoxidation, is very potent in lowering the transition temperature and is used in steels that are made specifically for low-temperature service.

Commercial mild steels, even those within the same grade, may exhibit considerable differences in their notch-sensitivity characteristics. These differences, which are not predictable from the conventional chemical analyses, may be attributed to various factors such as steel-making practice, deoxidation, grain size, finishing temperature, thickness of section, heat treatment, and others.

In the fully quenched and tempered condition, composition, other than high-carbon content, does not appear to be much of a factor provided the steels harden fully on quenching. This does not appear to hold true in the normalized or annealed conditions more commonly applied to piping materials. Steels that are susceptible to temper embrittlement are also exceptions.

Effect of composition on transition temperature of steel is still the subject of considerable study. There is some indication that nitrogen present as aluminum nitride may be beneficial, but detrimental if present as other nitrides.

Homogeneity. It is generally recognized that the concentration of carbon, alloying elements and impurities varies throughout the ingot and, consequently, throughout the finished steel plate or pipe. These variations in composition affect the conventional mechanical properties as well as the notch characteristics and the transition temperature. For example, the ductility in the tension test may increase, whereas the strength may decrease from the top to the bottom and from the center to the edge of the ingot from which plate or pipe are subsequently produced. In the Charpy keyhole and V-notch impact tests, the toughness may increase and the transition temperature may decrease from the top to the bottom of an ingot. These effects all were consistent with chemical variations.

178 *Defects and Failures in Pressure Vessels and Piping*

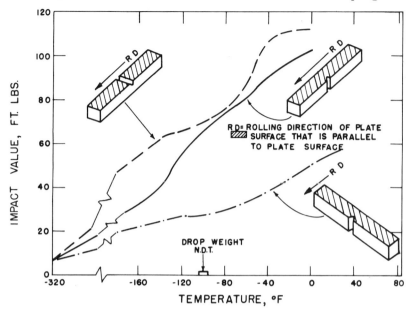

Fig. 9-19. Effect of rolling direction of plate on Charpy V-notch impact strength of one heat of 1-in. thick HY-80 steel.[34]

Some plate and pipe steels show marked directional properties such as higher values for ductility and notched-bar impact values when tests are made on specimens taken parallel to the direction of extruding or rolling as compared to specimens taken transverse to this direction. An example of the effects of rolling direction on the Charpy V-notch impact strength of 1-in. thick HY-80 steel plate is illustrated in Fig. 9-19.[34] On plate steels, directional properties sometimes may be minimized by cross rolling and by certain heating procedures prior to rolling. This is not possible on extruded pipe. An example of the effect of directionality in an ASTM A106 pipe is shown in Fig. 9-20 with Charpy V-notch impact test specimens removed parallel and perpendicular to the axis of the pipe.

Grain Size. The smaller the ferrite grain size, the lower generally will be the transition temperature, Fig. 9-21. Thus, the lower the final rolling temperature of the steel and the higher the cooling rate, the smaller will be the grain size and the lower will be the transition temperature unless the final rolling temperature is so low that the steel is to some degree cold rolled.

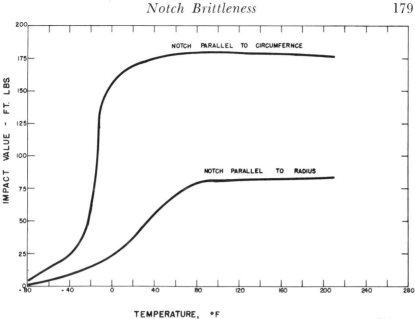

Fig. 9-20. Effect of directionality on Charpy V-notch impact test results on an ASTM A106 Grade B pipe section.

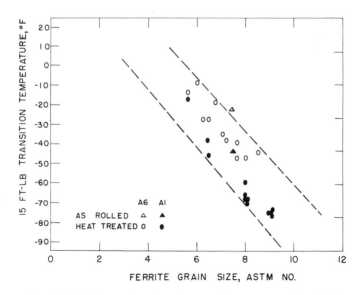

Fig. 9-21. Effect of grain size on Charpy V-notch 15-ft-lb transition temperature.[35]

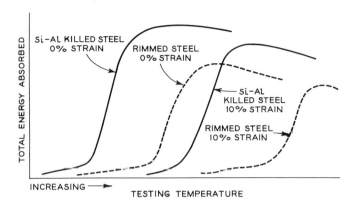

Fig. 9-22. Effects of straining on the transition temperatures of Si-Al killed and rimmed steels.

Aluminum and silicon additions during final deoxidation are useful in providing fine grain size in normalized material. A normalizing treatment is particularly effective if the pipe or plate was finished at a high rolling temperature.

Straining. Cold deformation or straining generally raises the transition temperature of a steel, Fig. 9-22, especially the nil-ductility transition. The fracture transition is not affected until the nil-ductility transition has been raised to the level of the fracture transition. This may have adverse effects on pipe which has been cold expanded or cold bent.

Charpy keyhole impact tests conducted on ASTM A201 (killed) plate and ASTM A70 (now A285, rimmed) plate have shown that a tensile strain of 1% in the rolling direction can raise the transition temperature by about 20°F for the killed steel and by about 50°F for the rimmed steel.[36] With increasing straining, the difference in effect on killed and rimmed steels tends to become less. Straining to 20% in the rolling direction can increase the transition range of both steels (A201 and A70) by about 80°F.[37]

However, it should be recognized that cold forming is extensively done on steel materials, which have provided satisfactory service. Examples represent cold expanded transmission piping, cold bending of pipe, cold rolling of vessel shells, and cold forming of pressure vessel heads.

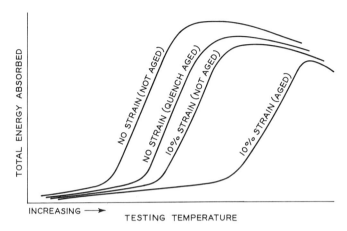

Fig. 9-23. Effects of straining and aging on the transition temperature of a rimmed steel.

Straining and Aging. The transition temperature of a steel, which is susceptible to strain aging, is raised to a higher temperature after straining and aging at 500 to 800°F than after straining alone.[36] This is illustrated in Fig. 9-23 which shows results which may be expected in a rimmed steel.

In many vessels and tanks, strain aging may occur localized near supports, attachment welds, welded baffles, etc.

Strain aging has been considered responsible for repeated failures in a catalytic cracker regenerator in an oil refinery.[45] Sudden 7-ft long and 2-ft long cracks developed in 1⅛-in. thick steel plate during shutdown periods after the regenerator cooled down. Cracking occurred in the areas where weirs had been attached by welding causing significant straining. During service at intermediate temperatures, aging had occurred causing the increase in the transition temperature illustrated in Fig. 9-24. Subsequent failures were avoided by design changes and the maintenance of a "temperature blanket" during shutdown periods at 175 to 210°F in the vessel area susceptible to strain-age embrittlement. In one instance, when a heavy rainstorm caused a failure of the heating elements, cracking occurred again.[45]

Heat Treatment. Some carbon steel plate or pipe is furnished in the hot-finished condition. Hot finishing is generally performed be-

182 *Defects and Failures in Pressure Vessels and Piping*

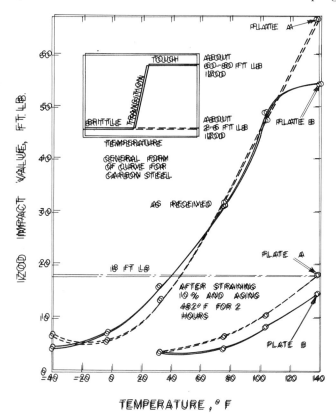

Fig. 9-24. Izod impact test results of catalytic cracker regenerator steel subject to strain-aging and cracking in service.[45]

tween 2200 and 1600°F. and is followed by air cooling. Hot pipe bends and extrusions are examples of hot-finished pipe products. Under these conditions, these steels can be compared to normalized steels, although it should be recognized that the temperature of finishing is an important factor.

As large sections are usually finished at higher temperatures and cool more slowly than the light sections, they exhibit higher transition temperatures. This is at least in part due to the coarser grain size, and to the coarseness and distribution of pearlite.

The form and distribution of the carbide have a profound influence on the transition temperature of steel. In general, lowest transition temperatures are obtained with liquid quenching and tempering heat treatments. Normalizing or normalizing and tempering is the second most effective treatment. Conventional annealing cycles consisting of heating to above the critical range followed by slow cooling to ambient temperatures result in high transition temperatures, especially if the annealing temperature is so high as to cause grain coarsening. Such treatments are not recommended for pipe for low-temperature service. Normalizing at about 1650°F and tempering or a stress-relief heat treatment at 1100°F or above, but not higher than 1250 to 1350°F are the preferred treatments. The range 1350 to 1450°F sometimes raises transition temperatures and should be avoided; likewise, low-temperature stress-relief heat treatments (below 900°F) are to be avoided in some instances, as they raise the transition temperatures.

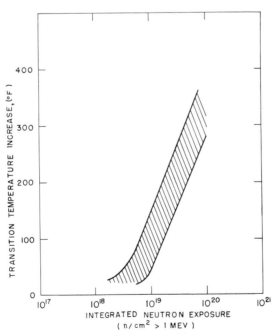

Fig. 9-25. Increase in transition temperature from neutron bombardment. Scatter band for ASTM A-212B, A-302B and A-336 steel irradiated at 500-575°F.[19]

Neutron Embrittlement. In light water and heavy water nuclear reactors, the neutron irradiation produced is generally of sufficient magnitude to raise significantly the transition temperature. The integrated fast neutron flux (>1 MeV) in the pressure vessel wall of a typical reactor usually varies between 4×10^{18} to 2×10^{19} n/cm². This neutron flux level may increase the Charpy V-notch transition temperature by 30 to 350°F, Fig. 9-25. For a specific irradiated nuclear pressure vessel, the embrittlement depends on the neutron dose, neutron spectrum, irradiation temperature, steel material, etc.[19, 38-41]

In nuclear reactor operations, consideration must be given to the possible shift in NDT due to radiation as well as the presence of stresses and defects in the vessel materials and welds. The combination of high stress and high neutron exposure may be most severe in the region of the vessel opposite to the center line of the core, particularly the longitudinal welds.[39] In addition, areas not so directly exposed to neutrons such as nozzle joints also represent potential hazards for brittle failures because of high stress concentration factors. Prescribed operating limits for several nuclear reactors are illustrated in Fig. 9-26.[39]

Mechanical Factors

Various mechanical factors may also exert considerable influence on the transition temperature of a steel. Most important are the stress system, section size, design, workmanship, welding practice, and others.

Stress System. Multiaxial tensile stresses raise the transition temperature. This is particularly true at the base of a notch or crack where multiaxial tensile stresses of considerable magnitude may develop.

Section Size. If the section size is increased without other changes in geometry, the transition temperature will also be increased. As is illustrated in Fig. 9-27, the amount of energy absorbed rises as the section size is first increased and the fracture remains ductile, but beyond a certain critical size, there is a sharp drop in energy accompanied by a change to brittle fracture. Further increase in size results in a very slight increase in energy absorption, but the fracture remains brittle. These results are partially due to the greater restraint of the wider specimens which increases the severity of the triaxial

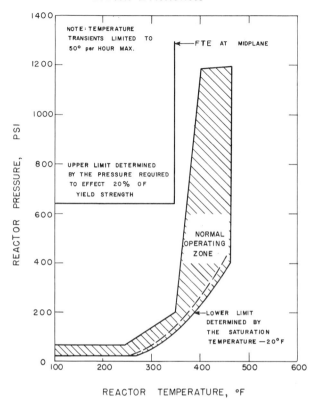

Fig. 9-26. Prescribed operating limits for several nuclear reactors.[39]

stresses. The standard size Charpy keyhole or V-notch impact specimen requires that the plate or pipe to be tested has a wall thickness of at least 2.5 mm. For thinner pipe walls, reduced section impact tests are frequently made. Typical test results on A106 pipe are illustrated in Fig. 9-28 involving full, 3/4, and 1/2 size Charpy V-notch specimens.

Design. Design based upon conventional tensile test data gives no assurance that a pressure vessel, tank or pipe section will not fail in a brittle manner. Nor can such assurance be obtained by simply increasing the section size with the intent of increasing the "factor of safety." In the presence of notches, increase in section size will most likely increase restraint and may even lead to failure at lower applied loads.

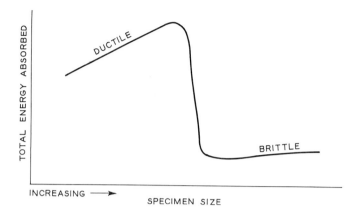

Fig. 9-27. Relationship between size of specimen and type of fracture.

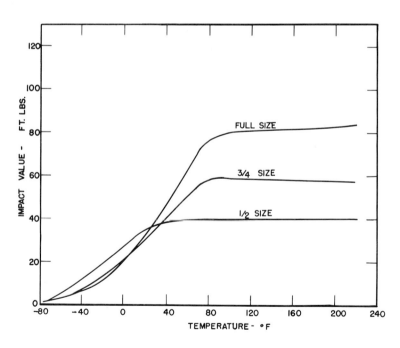

Fig. 9-28. Effect of specimen size on Charpy V-notch test results on ASTM A106, Grade B pipe material.

The designer must take into account the conditions which promote notch sensitivity and attempt to control these conditions so that the ductility transition temperature of the piping, under the particular conditions of design and service, will be below the temperature at which it operates. This may be accomplished by selecting the proper steel, by using generous radii, by limiting the extent of surface defects and by controlling welding and other fabrication procedures. When it is necessary to incorporate undesirable design features, or when economic considerations restrict selection of the steel, the adverse factors may be at least partly offset by using a low design stress and, where possible, avoiding shock conditions in service. It also can be assumed that if piping fails under these conditions, it will fail in a brittle manner. The hazards associated with sudden failure of piping will dictate whether it should be used. If these hazards are great enough, the fracture transition temperature should be below the minimum service temperature.

Workmanship and Welding. Workmanship represents an intangible but highly important factor, the effect of which is difficult to evaluate in the completed vessel, piping component or system. Inspection utilizing radiographic or ultrasonic methods or magnetic particle or liquid penetrant techniques at critical locations, have generally been helpful, in judging and controlling the quality of workmanship. These inspection procedures, when judiciously applied, also have the psychological effect of influencing the welder to improve his workmanship and thus reduce the probability of incorporating notches into the structure.

It is even more difficult to evaluate the brittle behavior of weld deposits than to evaluate the properties of wrought plate and seamless piping. This is so primarily because the conventional testing procedures have not been correlated with the service behavior of the material after it has been welded. For example, one electrode may be preferred over another because the weld metal produced from it develops higher tensile strength. However, it might well be that an incipient surface crack will propagate readily in the weld deposit made from the one electrode but not in the deposit from the other, because the latter has ability to adjust itself plastically at the service temperature.

In welded vessels and piping involving service where notch-brittle

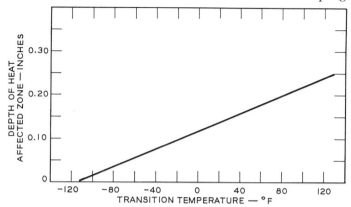

Fig. 9-29. Effect of depth of heat-affected zone on the transition temperature of a welded rimmed steel. (Composition: 0.16 C, 0.45 Mn, 0.01 Si)

behavior is a consideration, it is advisable to keep the width of the heat-affected zone as narrow as possible by means of rapid rates of electrode travel. When the layer thickness of the weld metal is held constant, faster rates of electrode travel reduce the width of the heat-affected zone. Apparently, the effect of the heat from the retreating arc is minimized, thus reducing the width of the heat-affected zone. This is desirable because transition temperatures tend to rise as the heat-affected zone becomes wider. The effects of width of the heat-affected zone on the transition temperature are illustrated in Fig. 9-29.

Welded structures in the as-welded state always contain residual welding stresses, which in certain sections may be as high as the yield strength of the steels. Under certain conditions, these residual stresses may not be particularly harmful. This is confirmed by the fact that many structures, piping and vessels have given highly satisfactory service without having received stress-relieving treatments subsequent to welding operations. Moreover, when failures have occurred, it was generally found that they were brought on by causes other than residual stresses alone.

When residual stresses are believed to be a factor in the service life of a vessel, tank or piping, a stress-relief heat treatment should be applied. In some instances, however, stress relieving may raise the transition temperature.

It has been demonstrated by certain restraint tests, such as the circular-groove tests, that for specific preheat and interpass temperatures, the degree of restraint during the cooling cycle affects the transition temperature of weld deposits. When controlling preheat and interpass temperatures at 70 to 100°F, the transition temperature of low-alloy steel weld metal deposits has been raised from -90 to $+10°F$ by increase in restraint. Where the degree of restraint is held constant, increasing the preheat and interpass temperatures effectively lowers the transition temperature of the weld deposit.

Welding at subfreezing temperatures is not permitted by most codes. Such practice will produce weld deposits having reduced ductility and impact toughness. The drop in ductility becomes pronounced at welding temperatures below $-50°F$. Results of impact tests showed that specimens welded at $-50°F$ exhibited a toughness 15 or 20% lower than specimens welded at ordinary atmospheric temperatures (75°F).

Extensive investigations of the last few years indicate that, due to a combination of hydrogen and straining, ordinary electrodes of the E6010 type may produce microcracks as the weld cools from 500°F to ambient temperature, particularly when welding heavy sections of plate or pipe, or when welding under severe restraint. (See Chapter 12). Preheating, or the use of E7015, E6015, E6018, E7016 or E7018 type electrodes circumvents this difficulty.

SERVICE IN THE BRITTLE TEMPERATURE RANGE

In a significant number of applications, pressure vessels, tanks and piping are exposed to temperatures below the Charpy V-notch transition temperature of the steels used. This may involve intermittent exposure, as during the shutdown of equipment operating at elevated temperatures, or may involve prolonged or continuous operating cycles at atmospheres or lower temperatures. Examples are structures exposed to severe climatic winter environments, refrigeration industry practice, production of carbon dioxide, and low-temperature gas separation.

There is no substantiated record of brittle service failures which have been both initiated and propagated in a brittle manner in stress-relieved and proof-tested vessels.[1, 3, 42, 43] In some stress-relieved vessels fractures resulting from proof testing were caused by cracks present in the as-welded vessel prior to the stress relieving heat treatment.[3] The failures which have occurred have been propagated in a brittle manner after having once started as a fatigue failure, or a corrosion crack. Such cracking generally has started at a weld defect, design notch or some other type of mechanical notch, as illustrated in Chapter 7.

A substantial number of brittle failures, however, have occurred in field-erected non-stress relieved tanks and vessels, and some in piping.

Brittle failures have occurred also during proof testing where the general stress level at the instant of failure has been of the same order of magnitude, or above, the yield strength of the material. During proof testing, residual stresses and local stress concentrations emanating from weld cracks, design notches, or other defects will be reduced. Without or during proof testing, these defects might act as initiating points for brittle failure.[44]

Because of a brittle failure during proof testing at 5300 psi of a $8\frac{1}{2}$-in. thick high-pressure heater, resulting in the loss of a life, one manufacturer has recommended that similar heaters be tested at a minimum water temperature of 100°F. Associated with the failure were high triaxial stresses at the point where the pass partition plate was joined to the tube sheet and channel wall.

In reviewing extensive brittle failure experience, Wells [43] has concluded that under certain conditions, brittle fracture will almost certainly not occur in the operating range of loading irrespective of operating temperature and to a limit approaching 180°F below the transition temperature. Such conditions are met (1) in vessels or piping without well rounded (design) details, but with an absence of flaws or (2) with an average rounding of (design) details, but in the presence of fabrication flaws, provided that the residual stresses and local embrittlement arising during fabrication have been removed by a furnace stress relieving operation.

Brittle fracture, however, will be a hazard under these conditions during fabrication or hydrostatic pressure testing. It will also become a hazard if welding or localized concentrated heating (for straightening) are done after stress relieving.

REFERENCES

1. Shank, M. E., "A Critical Survey of Brittle Fracture in Carbon Plate Steel Structures Other Than Ships," Welding Research Council Bulletin No. 17 (Jan. 1954).
2. Parker, E. R., "Brittle Fracture of Engineering Structures," New York, Wiley, Chapman & Hall, 1957.
3. Shank, M. E., "Control of Steel Construction to Avoid Brittle Failure," Welding Research Council, New York, 1957.
4. Biggs, W. D., "The Brittle Fracture of Steel," New York, Pitman Publishing Corp., 1960.
5. Averbach, B. L. et al. (Editors), "Fracture," New York, John Wiley & Sons, Inc., 1959.
6. Tipper, C. F., "The Brittle Fracture Story," Cambridge University Press, 1962.
7. Brown, A. L. and Smith, J. B., "Failure of Spherical Hydrogen Storage Tank," *Welding J.*, 24, 235–240 (Mar. 1945).
8. Brown, A. L. and Smith, J. B., "Failure of Spherical Hydrogen Storage Tank," *Mech. Eng.*, 66, 392–397 (June 6, 1944).
9. Plummer, F. L., "Field-Erected Pressure Vessels," *Welding J.*, 25, 1081–1089 (Nov. 1946).
10. Elliott, M. A., Seibel, C. W., Brown, F. W., Art, R. T., and Berger, L. B., "Report on the Investigation of the Fire at the Liquefication, Storage and Regasification Plant of the East Ohio Gas Co., Cleveland, Ohio," U.S. Dept. Interior, Bur. Mines, R. I. 3867 (Feb. 1946).
11. Jackson, J. O., "Liquefied Gas Storage Containers," *Gas Age*, 37–42 (Apr. 22, 1943).
12. Tourret, J., "Fissuration du Reservoir a 8 de Brut a la Raffinerie de Normandie," Compagnie Francaise de Raffinage, Direction Recherches et Procedes Rapport No. 1152 (July 23, 1951).
13. Oldacre, M. S., Chicago Utilities Research Commission Report (May 1, 1953).
14. Feely, F. J. and Northrup, M. S., "Failure of Two Oil Storage Tanks, Fawley, England," Standard Oil Development Co., Esso Engineering Dept. (Nov. 3, 1952); "Why Storage Tanks Fail," *Oil Gas J.*, 52, 73–77 (Feb. 1954).
15. Barr, W., "Failure of Welded Oil Storage Tanks," Motherwell, Scotland, Colvilles, Ltd., Mar. 23, 1953.
16. Robertson, T. S., "Propagation of Brittle Fracture in Steel," *J. Iron Steel Inst.*, 175, 361–374 (1953).
17. McClure, G. M., Eiber, R. J., Hahn, G. T., Boulger, F. W., and Masubuchi, K., "Research on the Properties of Line Pipe," Summary Report Catalog No. 40/PR, 1962, American Gas Association, New York.
18. Kerr, J. G., "Charpy Test Correlations from Unusual Pipeline Failures," *Welding J.*, 41, Res. Suppl., 257s–264s (1962).
19. Pellini, W. S., Steele, L. E., and Hawthorne, J. R., "Analysis of Engineering and Basic Research Aspects of Neutron Embrittlement of Steels," *Welding J.*, 41, Research Suppl., 455s–469s (1962).

20. Pellini, W. S. and Puzak, P. P., "Practical Considerations in Applying Laboratory Fracture Test Criteria to the Fracture-Safe Design of Pressure Vessels," *Trans. ASME,* **86,** Series A, 429–443 (1964).
21. Kihara, H. and Masubuchi, K., "Effect of Residual Stress on Brittle Fracture—Studies on Brittle Fracture of Welded Structure at Low Stress Level," Report of Transportation Technical Research Institute, Japan, No. 30 (1958); reprinted in *Welding Research Abroad,* **4,** 20–48 (Nov. 1958).
22. Pellini, W. S. and Puzak, P. P., "Fracture Analysis Diagram Procedures for the Fracture-Safe Engineering Design of Steel Structures," Welding Research Council Bulletin No. 88 (May 1963).
23. Kooistra, L. F., Lange, E. A., and Pickett, A. G., "Full-Size Pressure Vessel Testing and Its Application to Design," *Trans. ASME,* **86,** Series A, 419–428 (1964).
24. Nichols, R. W. and Harris, D. R., "Brittle Fracture and Irradiation Effects in Ferritic Pressure Vessel Steels," *ASTM Spec. Tech. Publ. No. 341,* 162–198 (Oct. 1962).
25. Kehoe, R. B. and Nichols, R. W., "Brittle Fracture and Reactor Pressure Circuits," *Nucl. Eng.,* **6,** 112 (1961).
26. Cowin, A. and Nichols, R. W., "Assessment of Steels for Nuclear Reactor Pressure Vessels," *Trans. ASME,* **86,** Series A, 393–402 (1964).
27. Fearnehough, G. D. and Vaughan, H. G., "Comparisons Between Drop Weight and Crack Arrest Tests for the Estimation of the Brittle Fracture Transition Temperature of Steel," *Welding J.,* **42,** Res. Suppl., 202s–204s (1963).
28. Wright, D. J. et al., "The Selection of Steel for Notch Toughness," in "Metals Handbook," 8th Ed., Vol. I, pp. 225–243, American Society for Metals, Metals Park, Ohio, 1961.
29. Puzak, P. P. and Pellini, W. S., "Evaluation of Notch-Bend Specimens," *Welding J.,* **33,** Res. Suppl., 187s–192s (1954).
30. Pellini, W. S. and Eschbacher, E. W., "Ductility Transition of Weld Metal," *Welding J.,* **33,** Res. Suppl., 16s–20s (1954).
31. ASTM Specification E208-63T, "Tentative Method for Conducting Drop-Weight Test to Determine Nil-Ductility Transition Temperature of Ferritic Steels."
32. Feely, F. J., Jr., Hrtko, D., Kleppe, S. R., and Northup, M. S., "Report on Brittle Fractures Studies," *Welding J.,* **33,** Res. Suppl., 99s–111s (1954).
33. Feely, F. J., Jr., Northup, M. S., Kleppe, S. R., and Gensamer, M., "Studies on the Brittle Failure of Tankage Steel Plates," *Welding J.,* **34,** Res. Suppl., 596s–607s (1955).
34. Sibley, C. R., "Welding of HY-80 Steel with the Gas-Shielded Metal Arc Process," *Welding J.,* **42,** Res. Suppl., 219s–232s (1963).
35. Kihara, H., Suzuki, H., and Tamura, H., "Researches on Weldable High Strength Welds," Soc. of Naval Architects, Tokyo, Japan (1957).
36. Osborn, C. J., Scotchbrook, A. F., Stout, R. D., and Johnston, B. G., "Effect of Plastic Strain and Heat Treatment," *Welding J.,* **28,** Res. Suppl., 337s–353s (1949).
37. Tör, S. S., Stout, R. D., and Johnston, B. G., "Further Tests on Effects of Plastic Strain and Heat Treatment," *Welding J.,* **30,** Res. Suppl., 576s–583s (1951).

38. Symposium on Steels for Reactor Pressure Circuits, British Iron and Steel Institute, Special Report No. 69, London (1961).
39. DiNunno, J. J. and Holt, A. B., "Radiation Embrittlement of Reactor Vessels," *Nucl. Safety*, **4**, 34–47 (Dec. 1962).
40. McLaughlin, D. W., "Neutrons Embrittle Reactor Vessels of Carbon Steel," *Nucleonics*, **21**, 36–41 (Feb. 1963).
41. Nichols, R. W., "Effect of Irradiation on the Properties of Steels," Welding of Power Plants, 1964 IIW Annual Assembly, Prague, Czechoslovakia.
42. Puzak, P. P., Babecki, A. J., and Pellini, W. S., "Correlations of Brittle-Fracture Service Failures With Laboratory Notch-Ductility Tests," *Welding J.*, **37**, Res. Suppl., 391s–407s (1958).
43. Wells, A. A., "Pressure Vessel Brittle Fracture and Information Required to Design Against It," *Welding Research Abroad*, **6**, 85–93 (Nov. 1960).
44. Wallin, L., "Brittle Fracture in Pressure Vessels," Welding Research Abroad, **10**, 49–65 (Feb. 1964).
45. Pull, D. J., "Mechanical Engineering in a Modern Oil Refinery," *Proc. Inst. Mech. Eng.*, **176**, 495–521 (1962).

———chapter 10

FABRICATION DEFECTS

Pressure vessels and piping systems normally are fabricated, welded and inspected under applicable established codes.[1, 2, 3] In spite of apparent compliance, failures occur and the resulting losses in production frequently cost far more than the pressure vessel or piping component that failed. In a modern steam power plant, a one-day shutdown of a boiler unit may involve a loss exceeding $15,000. In chemical plants and refineries, failures of critical equipment or piping components may be even more costly.

Because of the importance of sound fabrication and the many different types of failures resulting from the lack of it, these failures are discussed in separate chapters.

Chapters 10 to 15 deal with a portion of the failures that can be attributed primarily to improper fabrication, either in shop fabrication plants or during field erection. The direct causes of these types of failure can be further separated into several groupings:

Chapter 10—shop forming, bending and working
Chapter 11—end preparation for welding
Chapter 13—welding
Chapter 14—heat treatment
Chapter 15—cleaning

Chapter 12 discusses weld defects which may or may not be critical.

EXPERIENCE OF FABRICATOR

Not enough can be said about the importance of know-how, experience, and responsibility on the part of the fabricator. Many failures are the direct result of the lack of one or all of these qualifications. Too many fabricators bid on pressure vessel, tank and piping jobs for which they lack the personnel experienced in welding, metallurgy, quality control and inspection. It is also essential that the fabricator understands fully the applicable codes and specifications, and is familiar with the service requirements.

Fabricators, consulting engineers, and purchasers share alike in the blame for the defective workmanship that occurs; fabricators for overeagerness to expand and excessive price cutting; consulting engineers for carelessly prepared specifications; and purchasers for failure to check specifications, lack of appreciation of proper quality, and misplaced cost consciousness.

Fabrication of pressure vessels includes rolling or press-braking of plates, bending of component parts, etc.

Fabrication of piping may involve hot or cold bending, hot or cold extruding, hot swedging and Van Stoning.

The procedures for these fabrication operations must take into account the characteristics of the equipment involved and the forming properties of the specific plate, pipe or tube materials.

PIPE BENDING

Pipe sections are often bent arbitrarily to a five diameter radius. However, sharper radii of three pipe diameters sometimes are required. On the other hand, bending radii representing six, ten or fifteen times the pipe diameter may be desired.

Cold Bending

Cold bending is normally done on tubing and on pipe with relatively small diameters and wall thicknesses. In some plants, however, equipment is available for bending pipe with outside diameters of $10\tfrac{3}{4}$-in. and in Schedules to 160, and in light wall thicknesses up to 24 in. OD. Proper equipment and dies are essential to avoid wrinkling, excessive thinning, and excessive ovality. In general,

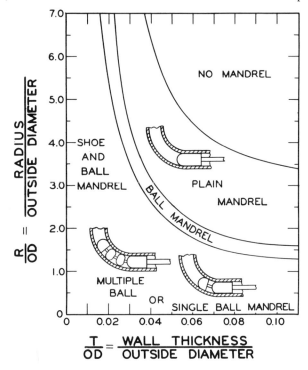

Fig. 10-1. Sketch illustrating equipment requirements for proper cold bending of pipe.

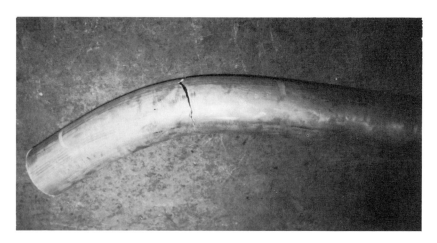

Fig. 10-2. Cracking in a 4½-in. OD by 0.215-in. wall 6063-T6 aluminum pipe during cold bending to a five diameter radius.

Fabrication Defects

greater care is required with decreasing wall thicknesses and smaller bending radii. Figure 10-1 illustrates the equipment requirements considered normal for proper cold bending.

Cracking in a 4½-in. OD by 0.215-in. wall 6063-T6 aluminum pipe bend is illustrated in Fig. 10-2. Lack of a mandrel support resulting in uneven straining was considered responsible for the failure.

In some materials, the bending characteristics may depend on prior thermal history. For example, certain alloy steels normally quenched from above 1600°F and then tempered may fail during subsequent bending when the temperatures of the steel after quenching prior to tempering did not fall below 400°F. Brittle failures which have occurred have been associated with localized hardening of 280 to 322 Brinell.[4] In these materials, it may sometimes be desirable to check the proper heat treatment by hardness and Charpy V-notch impact tests.

Hot Bending

Hot bending is done at elevated temperatures. On carbon, alloy and stainless steels, the desirable temperature range is normally between approximately 2000 and 1500°F. The lower limit depends on the particular alloy steel involved. Except for very heavy wall thicknesses, sand filling and tight packing are generally essential to avoid excessive ovality and wrinkling.

In some instances, cracking may form during hot bending. Sometimes, this may be the result of internal defects in the material. It may also result from an inherent lack of ductility (hot shortness) at elevated temperatures (Figs. 7-25 and 7-26).

Although cracks formed during hot bending may reach the inside surface of pipe, they may not be detected because of the relative inaccessibility of the inside of the pipe. Figure 10-3 illustrates a service failure due to cracking in a 12Cr–1Mo–0.4W–0.3V bent steel pipe section 12½-in. OD by ½-in. wall connecting a steam turbine to the intermediate superheater.[5] At the time of failure, the crack measured 4-in. on the inside and ⅞-in. on the outside. Figure 10-4 shows a cross section through part of the cracked area at magnifications of 1X and 500X. The decarburization and ferrite along the crack edges indicated that the cracking was present after hot bending prior to annealing.

Fig. 10-3. Cracking on inside of pipe bend after 4000 hrs. at 1040°F 926 psi service.[5]

Cracking in some alloy steel piping has also been caused when, during hot bending, water is applied locally for a prolonged period to set the particular pipe area and avoid excessive thinning. Fracture in a 5Cr–½Mo 6-in. OD by 0.432-in. wall pipe has been reported. The cracking occurred in the martensitic structure produced at a location corresponding to the original austenite grain boundaries.[6]

Improper hot bending and subsequent local heating were primarily considered responsible for the failure of a 14-in. OD by 0.56-in. wall pipe.[7] Failure of the gas-transmission pipe occurred after less than 100 hours of service at 1800 psi, about 200 psi below the design pressure, and 1450 psi below the hydrostatic test pressure. The pipe temperatures at the time of failure varied between 125 and 150°F. The pipe fractured into 88 pieces, some thrown as far as 1500 ft. The pipe was seamless, exhibiting a yield strength range from 55,500 to 66,000 psi and elongation values between 33 and 40% in 2 in. The average chemical composition was 0.26C, 0.84Mn, 0.30Si, 0.026P, and 0.036S. Examination of the microstructure revealed untempered martensite and tempered martensite. The bends had been hot formed at 1800°F. Water quenching was utilized during hot bending. Subsequently,

Fabrication Defects 199

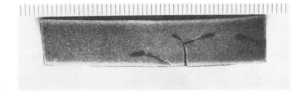

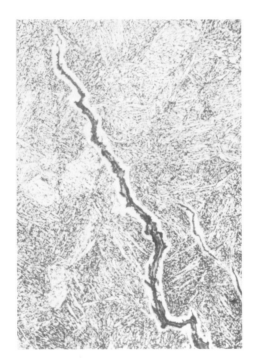

Fig. 10-4. Cross sections of cracked area illustrating decarburization in steel along sides of crack.[5] (Magnifications 1× and 500×)

surface irregularities were removed by local torch heating, hammering and water quenching.[7]

Improper heating, sand filling or nonuniform bending may result in wrinkles, buckles or excessive ovality. These problems increase with increasing diameters and decreasing wall thickness.

Hammering to remove wrinkles or pressing to reduce ovality are methods used to improve the contour and uniformity of the bends.

If improperly done, these operations, although improving the appearance of the bends, may result in greater damage. Service failures may then be the consequence.

For example, a 12-in. long crack occurred after 3000 hours along the neutral axis of a pipe bend in a 10½-in. OD by 1.06-in. wall high temperature steam piping system of 1Cr–½Mo steel. At the crack location, the ovality of the bend was 7.1%. However, the larger diameter was in the plane of the bend, and not perpendicular as is normally the case in hot bends. Along the neutral axis, pressure marks were also apparent, indicating that the pipe bend had been subjected to a pressing operation after hot bending. The apparently severe cold pressing operation, which was not followed by a subsequent stress relief was considered the primary condition leading to the service failure.[8]

Sand left in hot bent boiler tubing has resulted in tube ruptures and even in bursting.[9] The sand restricted circulation and resulted in severe overheating.

Cracking has occurred readily in nickel and high-nickel alloy pipe and tubing when hot bending was done with sand containing sulfur and sulfur compouds. Even sand containing as low as 0.012% sulfur has caused hot cracking during bending.[10] Washed and dried high-silica sand should be used. Burning out of residual sulfur by heating thin layers of sand in an oxidizing atmosphere at 2100°F may also provide satisfactory sand conditions.[10]

Ovality

Excessive ovality in cold bent boiler tubing has resulted in a number of failures caused by corrosion fatigue (see Chapter 17). Maximum stresses develop along the neutral axis. In Germany,[8] the maximum ovality permissible in tube and pipe bends is determined on the basis of outside diameter and wall thickness as follows:

$\frac{OD}{ID}$ Ratio	1.1	1.15	1.2	1.25	1.3	1.35	1.4
Max. Ovality, %	3.5	5.25	7.25	9	11	12.75	14.75

In American practice, the maximum permissible ovality is considered to be 8%.[11]

The lighter the wall thickness, the greater tends to be the ovality.

EXTRUDING OF OUTLETS

The extrusion of outlets calls for substantial working and requires an understanding of the physical properties and mechanical characteristics of the piping material involved. Figure 10-5 illustrates cracking in an extruded outlet on a Type 304 stainless steel pipe containing a high percentage of ferrite in the normally austenitic metallurgical microstructure. Normally, Type 304 stainless steel materials contain between 0 and 8% ferrite in the austenite matrix. However, the stainless steel shown in Figs. 10-5 and 10-6 was modified chemically by increasing chromium and decreasing nickel to near the specified limits to produce about 15 to 20% ferrite resulting in increased tensile and yield strength properties. In turn, hot ductility was reduced so that the alloy no longer was suitable for severe hot working or forming operations. However, even commercial wrought carbon steel pipe made to ASTM Specification A106 occasionally may be unsuitable for the extrusion of outlets, as is illustrated in Fig. 10-7.

Fig. 10-5. Cracking of extruded outlet in type 304 stainless steel pipe 10-in. OD x 1⅝-in. wall.

202   *Defects and Failures in Pressure Vessels and Piping*

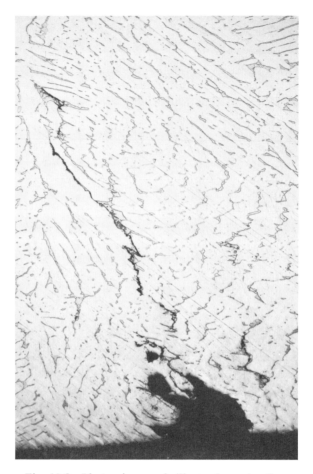

Fig. 10-6. Photomicrograph illustrating microfissuring along ferrite component in centrifugally cast Type 304 stainless steel pipe. (Magnification 50×)

SWEDGING OF REDUCING ENDS

Swedging involves the reduction of a pipe end to a smaller diameter as shown in Fig. 10-8. On most piping materials, reductions of more than 50% can be satisfactorily made. The reductions in diameter that can be made in each pass, however, vary with the material. Excessive reductions may produce cracking on the inside of the pipe. This is illustrated in Fig. 10-9.

Fabrication Defects 203

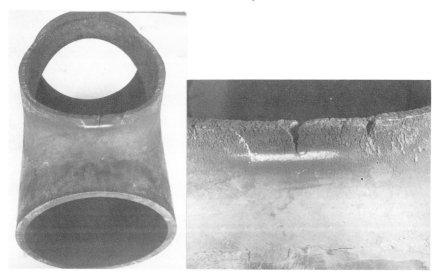

Fig. 10-7. Cracking of extruded outlet in ASTM A106 Grade B carbon steel pipe 14 in. OD x ⅜-in. wall.

Fig. 10-8. Swedging of carbon steel pipe end.

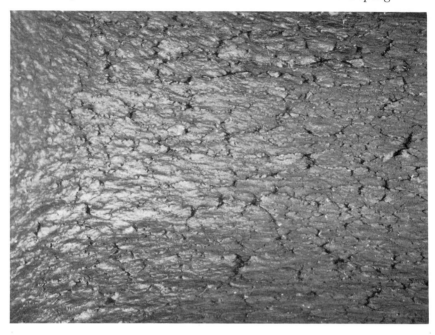

Fig. 10-9. Cracking on inside of improperly swedged end in 8⅝-in. OD x 1-in. wall Type 304 stainless steel pipe.

REFERENCES

1. Boiler and Pressure Vessel Code, American Society of Mechanical Engineers, New York, N.Y.
2. American Standard Code for Pressure Piping, American Society of Mechanical Engineers, New York, N.Y.
3. Standard 1104 for Field Welding of Pipelines, American Petroleum Institute, New York, N.Y.
4. Verein. d. Grosskesselbesitzer 1960/61 Review.
5. Private Communication, Erdoelchemie, A. G., Germany.
6. Rutherford, J. J. B., "Some Experiences in Service (Power, Oil and Chemical Plants)," *High Temperature Properties of Metals,* The American Society for Metals, Cleveland, 1951.
7. Kerr, J. G., "Charpy Test Correlations from Unusual Pipeline Failure," *Welding J.,* **41,** Res. Suppl., 257s–264s (1962).
8. Verein. d. Grosskesselbesitzer 1959/60 Review.
9. Davis, R. F., "The Development of the Large Assisted-Circulation Boiler in England," *Proc. Inst. Mech. Eng.,* **177,** 537–569 (1963).
10. "Fabrication and Design of Nickel and High-Nickel Alloy Pipe and Tubing," Tech. Bull. T-17, The International Nickel Co., New York.
11. Pipe Fabrication Institute, Results of Cooperative Test Program of Member Companies, 1960–1963.

chapter 11

END PREPARATION AND FIT-UP FOR WELDING

PREPARATION OF WELDING BEVELS

The preparation of the plate or pipe ends for welding may involve machining operations or may be done by flame or arc cutting. Frequently, a plate or pipe is sectioned initially by flame cutting. The ends are then machined. Post mills are used extensively for machining of pipe ends, particularly in heavier wall thicknesses.

When machining is done, care must be taken that deep notches or grooves are not produced on the plate or on the inside of the pipe. If pipe ends are machined internally for better fit-up, the change of contour should be gradual (tapered) along the inside surface.

Particular care should be exercised when flame or arc cutting procedures are employed. Excessive localized heating can be harmful to the material involved. Special subsequent heat treatments may then be necessary to remove the harmful heating effects.

A failure by corrosion of a carbon steel pipe used in a boiler feed water piping system is shown in Fig. 11-1. Etching of a cross section, shown in Fig. 11-2, evidences a change in the metallurgical structure of the steel, indicating that the pipe had been heated for the purpose of making a cut. Examination of the metallurgical structure provided further evidence that the localized area had been heated to approximately 2200°F.

206   *Defects and Failures in Pressure Vessels and Piping*

Fig. 11-1. Corrosion failures in ASTM A106 Grade B pipe in boiler feed water piping system.

Fig. 11-2. Etching of cross section slicing across edge of failure evidences change in metallurgical structure caused by local heating to about 2200°F.

End Preparation and Fit-up for Welding 207

Fig. 11-3. Cracking resulting from poor fit-up producing notch between weld and header.

WELD JOINT FIT-UP

Good joint fit-up is essential to the making of a sound weld. Depending on the welding process used, a slight mismatch may be permissible. Joint fit-up considerations include the root spacing between the mating ends of the vessel or pipe sections. Tight butting, or an insufficient spacing, may result in lack of penetration notches.

Double Bevel Joints

Butt welds in pressure vessels, tanks and pipe in diameters of over 24 in. are generally welded from the outside and inside. This facilitates fit-up and permits finishing by chipping or grinding where undesirable surface notches have resulted. Another advantage is close surface inspection for obvious notch-type defects.

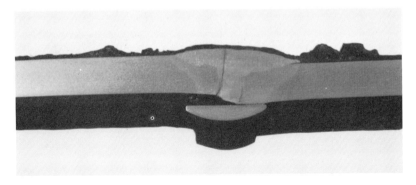

Fig. 11-4. Poor fit-up producing notches resulted in failure in 18-in. OD by 0.562-in. wall carbon steel pipe in process steam application.

That notches can result in failure is illustrated in Fig. 11-3 showing cracking adjacent to the weld between a 30-in. diameter, ½-in. thick steam header and the end cap. Failure occurred after 7500 hours and started at the notch formed between the header and weld. Thermal fatigue contributed to crack propagation.

Pipe Joints With Backing Rings

Even though a backing ring is used, poor fit-up may result in failures of the type illustrated in Fig. 11-4. The service involved steam in a process piping system. Notch conditions formed by the poor fit-up led to stress corrosion cracking.

Even if the fit-up is good, improper welding that results in a lack of penetration of the initial weld bead in the backing ring can cause failures in steam piping. Figure 11-5 illustrates a cracking failure

Fig. 11-5. Lack of penetration leaving weld crater in root of weld caused failure by stress corrosion cracking in 14-in. OD by 0.438-in. wall carbon steel pipe weld. Section of backing ring has been removed.

End Preparation and Fit-up for Welding

starting from the inside of the pipe at a weld crater. Stress corrosion cracking caused propagation of the crater notch, which finally resulted in failure. After the weld had failed, the backing ring was chipped out. The white area in Fig. 11-5 represents proper weld penetration into the backing ring.

If a weld has been properly made and the first pass accomplishes complete penetration into the backing ring, a failure can nevertheless occur if the backing ring has been tacked across the split. This is illustrated in Fig. 11-6. The split in the backing ring represents a notch condition which, in stress corrosive service environments, may propagate into cracking in the tack weld and continue from there into the pipe base metal. Ultimately, it may pass across the wall thickness. Had the tack weld been located away from the split or had the split been welded shut, the pipe base metal would not have been likely to fail.

Fig. 11-6. Split in backing ring provided notch into tack weld for initiation and propagation of stress corrosion cracking into 12-in. OD by 0.406-in. wall carbon steel pipe weld joint.

210 *Defects and Failures in Pressure Vessels and Piping*

Fig. 11-7. Lack of penetration in root of butt weld led to failure in this carbon steel boiler feed water pipe by stress corrosion cracking.

Pipe Joints Without Backing Rings

In pipe joints where the inside diameters have not been matched up by internal boring or tapering, fit-up may become a problem, particularly when backing rings are not used.

Figure 11-7 illustrates a failure in a weld in a boiler feed water piping system. Improper end preparations of the 5/8-in. thick pipe, lack of a backing ring, and careless fit-up and welding resulted in considerable lack of penetration in the root of the butt weld.

Socket Welds

Small diameter tubing connecting to pressure vessels and piping involving sampling, instrumentation or by-pass lines, thermometer wells, etc., usually involve socket-type joints. The small diameter tube is then joined to the vessel or pipe surface with a fillet weld.

Failures have resulted from so-called *fretting corrosion* (wear) or *friction oxidation* causing fatigue cracks in the surfaces which are only in "friction" contact with each other.[1]

Friction oxidation may be caused by mechanical vibrations, as for example, by electric fans.[2] The heat evolved from the friction between the two surfaces may be of such a magnitude that oxides formed at the surfaces readily cause and promote intergranular thermal fatigue cracking, Fig. 11-8.[2]

End Preparation and Fit-up for Welding

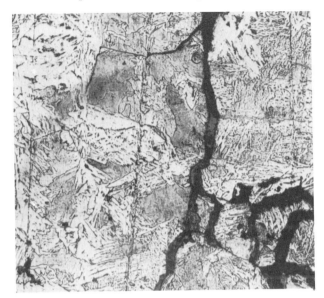

Fig. 11-8. Cracking and oxidation caused by friction contact in a 12% chromium stainless steel.[2] (Magnification 200×)

When the pipe or tube end is fitted tightly into the bottom of the socket, failure can occur in service involving thermal fatigue. Figure 11-9 illustrates cracking around the socket weld of a 1-in. diameter by-pass line into a 12-in. 1500-lb boiler stop valve in service at 950°F, 1200 psi service. The material was $1\frac{1}{4}$Cr–$\frac{1}{2}$Mo. During heating, the steam passing through the by-pass piping tended to heat the socket section to a higher temperature than the adjacent valve body. Because of the tight fit, the expansion of the pipe end was restrained resulting in a high stress in the socket weld and in ultimate failure.

To prevent these failures, it is usually good set-up practice to leave a space between the end of the tube and the vessel or pipe wall of $\frac{1}{16}$ in. Recommended set-up for socket and nozzle welds are illustrated in Fig. 11-10.* Another purpose of the mating metal surfaces is to prevent overstressing which may result from shrinkage of the fillet weld, particularly when making the connection shown in Fig. 11-10B.[3]

* Advantages or limitations of the joint designs shown are not under consideration here (see Chapter 2).

212 *Defects and Failures in Pressure Vessels and Piping*

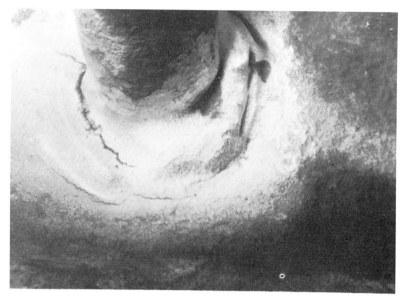

Fig. 11-9. Failure around socket weld between steam by-pass line and valve body caused by tight root fit and thermal fatigue.

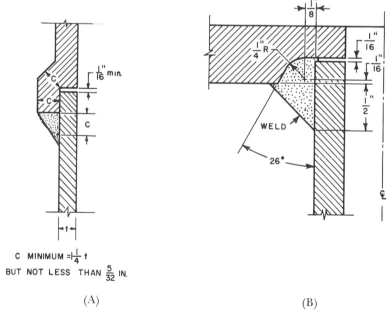

(A)   (B)

Fig. 11-10. Preferred pipe joint set-up involving connections of small diameter tubing to vessels and pipe. (A) Socket joints, (B) nozzle joints.

REFERENCES

1. Thielsch, H., "Thermal Fatigue and Shock," Welding Research Council Bulletin No. 10 (Apr. 1952).
2. Noren, T., Private Communication (see reference 1).
3. Weisberg, H., Private Communication (see reference 1).

chapter 12

WELD DEFECTS

CORRELATION OF TEST RESULTS TO SERVICE BEHAVIOR

As previously discussed in Chapter 7 on wrought plate and pipe materials, and Chapter 8 on castings, the term *defect* is broadly used. It may mean discontinuity, imperfection, flaw, or inhomogeneity. Defects in weld deposits may be of mechanical or metallurgical types.

Weld deposits, like base metals, are not perfect. This is recognized by all major welding codes which make allowances for the presence of defects in welds and set limitations on many defect conditions.

Some weld defects or inhomogeneities are extremely unlikely to result in a service failure. Others can be critical under specific service conditions or environments, and have caused or contributed to failures.

The effects of defects in welds and base materials on the service life of pressure vessels, tanks, and piping are largely misunderstood by engineers and in inspection codes establishing acceptability limits. Great emphasis and tight restrictions are placed on some defects, while others are ignored, even though they may be considerably more detrimental to the service life. Some so-called weld defects merely represent changes in the straightness or uniformity of the surface contour between the weld and base metal or within the weld.

Attempts to correlate published test results evaluating defect conditions with actual service experience often meet with difficulty, and may even be impossible. Many weldments which have provided completely satisfactory service under operating conditions considered to be extremely critical have contained defects of a magnitude claimed to be extremely dangerous in specific laboratory tests.

For example, in tension fatigue tests, cracking can be initiated readily in the weld ripple of a cover pass and be propagated into a failure. Similar or far more severe weld ripples have never caused failures in hundreds of thousands of critical pipe welds involving the major high-temperature high-pressure steam piping systems in power plants now in operation. Ordinary weld ripples should be considered no more hazardous than surface laps or slivers commonly found and accepted on pipe and plate materials.

As a most unfortunate consequence of the voluminous test literature on welds, engineers have become extensively concerned and preoccupied with minor weld defects and accept or ignore defects of much greater magnitude in castings, forgings, or plate products.

Nevertheless, there may be specific isolated service applications where even the most minute weld (or base) metal defect may lead to a failure or leak. Where such requirements apply, the laboratory results of specific tests or simulated service tests may then provide useful information.

Figure 12-1 illustrates a welded dome of Type 347 stainless steel for a missile application. Prior to machining, a 3-in. diameter 1-in. thick plug was welded into the $3/8$-in. thick dome. The weld had to meet normal levels of quality. The dome was subsequently machined to a 0.040-in. thickness and was again examined by X-ray radiography. At this thickness, weld defects of any kind were not permissible. Except for grain structure patterns in the forging base metal, actual weld defects were not apparent. Nevertheless, subsequent mass spectrometer testing showed, on one set of 6 specimens, minute leakage representing "hole" sizes of 0.00004-in. diameter. In a second lot of specimens, leaks were not detected. Subsequent analyses and laboratory tests showed that in the base metal along the edge of the welds in the defective lots, columbium compounds tended to form, resulting in microfissuring. Similar conditions leading to failures in Type 347 stainless steel base materials after prolonged service were discussed in Chapter 4. In the second lot of 13 leak-free specimens, this formation

216 *Defects and Failures in Pressure Vessels and Piping*

Fig. 12-1. Machined and welded Type 347 stainless steel forged dome containing minute flaws along weld edge. (Drawn lines locate weld area.)

was not observed. The only significant difference was in the forged base metals. The non-leaking grade contained 0.07% carbon, whereas the leaking grade contained 0.05% carbon. All other alloying elements were present in almost identical quantities. The columbium content was 0.72%. Hot bend tests indicated that the 0.05% carbon grade exhibited lower than normal hot ductility properties and a lower liquation temperature.

The point here is that laboratory tests were helpful (as well as in the Type 347 pipe materials discussed in Chapter 4) in explaining the conditions producing the defective materials and separating the acceptable from the rejectable materials.

On the other hand, it can be readily shown that a surface porosity

Weld Defects

of 0.04-in. diameter on a polished test specimen subjected to severe fatigue testing reduces fatigue life significantly. However, it is not realistic to equate this result to the average or even critical service environments to which welds in pressure vessels and piping are subjected.

In almost all cases, the respective authors of laboratory test studies do not themselves attempt such correlations. Unfortunately, engineers writing specifications often tend to make such erroneous interpretations.

Since, as stated above, test evaluations of specific weld defects may provide useful information for specific applications, the results and conclusions of some of the laboratory test studies of specific weld defects are reviewed in this Chapter. Actual service experience involving welds containing similar defects in pressure vessels and piping will also be recorded.

The reader must then form his own conclusions for the specific applications in which he is interested.

DEFECT CLASSIFICATIONS

Weld defects have been studied far more extensively than base-metal defects. Numerous research reports have been published evaluating, by detailed metallographic analyses and destructive tests, various types of weld defects. Almost any kind of defect has been proved hazardous under one or several sets of conditions. Yet, in other tests, the same defect may have been found to be of little or no consequence. Similar tests of several investigators may provide different results which have sometimes led to contradictory interpretations and conclusions.

ARC STRIKES

Definition

Arc strikes represent any localized heat-affected zone or change in the surface contour of the finished weld or base metal caused by an arc. Arc strikes can also be produced by heat generated by the passage of an electrical current, as may result from a welding electrode or from a magnetic particle inspection electrode. Sometimes a small

218 *Defects and Failures in Pressure Vessels and Piping*

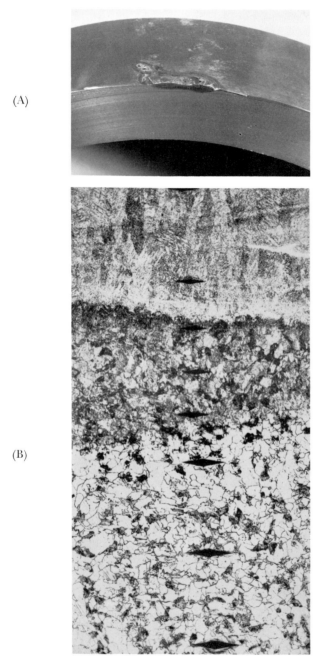

Fig. 12-2. Cross section through an arc strike in 2¼ Cr–1 Mo steel and sketch illustrating hardness values. [Magnifications—(A) 1×, (B) 60×] (Hardness values converted from Knoop scale to Brinell numbers.)

Weld Defects   219

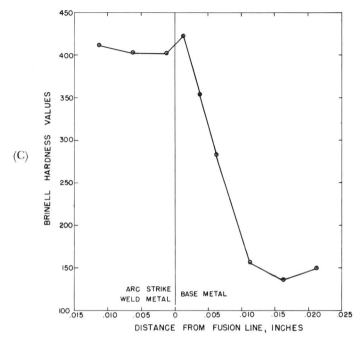

Fig. 12-2 cont'd

undercut may remain adjacent to an arc strike. Arc strikes also tend to produce hardening, particularly in the higher carbon steels and in alloy steels.

On the higher strength hardenable steel grades, arc strikes may also cause severe localized cracking.[1]

A typical arc strike is shown in Fig. 12-2.

Test Results

Under normal static mechanical fatigue test conditions, arc strikes have been considered of little consequence.[2] At high stresses and low number of cycles, arc strikes have not been critical. However, in reverse bend fatigue tests involving 5 million cycles, arc strikes have reduced fatigue strength in a low-carbon (0.04C) rimmed steel by 40%, and in a 0.26C–0.62Mn killed steel by 55%. The hardness in the heat-affected zone of the low-carbon steel was 285 Brinell.[3]

In mild and low-alloy steels, because of the severe hardening which is caused by arc strikes, the bend ductility can be significantly reduced, Fig. 12-3.[4, 5] Moreover, the transition temperatures may be increased.[4, 5]

In overpressure bursting tests and explosion bursting tests, failures often occur at arc strikes or tack welds.[6] Even where the arc strike "weld deposit" has been ground smooth, it still represents a metallurgical notch unless subsequent heat treatment reduces hardness differences.

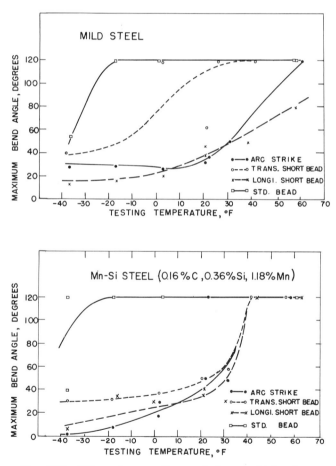

Fig. 12-3. Effect of arc strikes and short weld beads on (Kommerell) transition curves.[5]

Service Experience

Only rarely have arc strikes produced cracking leading to service failures. An example of cracking adjacent to an arc strike is shown in Fig. 12-4. The crack starting at the outside surface has gradually grown during service and has been filled with oxides as a result of service at elevated temperatures.

Arc strikes have also occasionally resulted in *underbead* cracking. In the majority of instances involving a $1\frac{1}{2}$Cr–$\frac{1}{2}$Mo and $2\frac{1}{4}$Cr–1Mo alloy steels, crack propagation during service has not occurred. The arc strikes were sufficiently close to the butt weld to be included in the stress relief heat treatments involving between 1 and 3 hours at temperatures between 1250 and 1375°F. This has removed the effects of localized hardening and residual stresses, and has improved toughness and ductility.

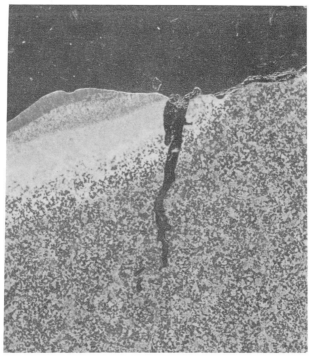

Fig. 12-4. Crack starting on pipe surface adjacent to arc strike on $\frac{1}{2}$ Mo steel. (Magnification 18×)

Nevertheless, some failures have been recorded. For example, cracking as a result of arc strikes on steels has been observed after several years of service.[7]

BACKING RINGS

Definition

A backing ring is a preformed metal strip extensively used in pipe welding to back up the joint during welding in order to facilitate a sound weld at the root.

Although not strictly a defect, backing rings are included in the discussion here as they represent a change in surface contour similar to a weld reinforcement.

Test Results

The results of bending fatigue tests on pipe butt welds welded from one side and made with flat backing rings are illustrated in Fig. 12-5.[8] The tests show a substantial reduction in fatigue life.

Cantilever-beam type fatigue tests on 2-in. and 4-in. Schedule 40 pipe test sections with flat backing rings have also produced a significant reduction in fatigue life.[9] However, when the backing ring was spaced $\frac{1}{16}$-in. below the pipe inside, leaving a $\frac{1}{16}$-in. gap between backing ring and inside pipe wall, the effect of the backing ring on fatigue life became inconsequential.[9]

Service Experience

Backing rings involving carbon steel, alloy steel, and aluminum materials have been used extensively in pipe butt welds for over 30 years. Their use has included critical service applications involving temperatures as high as 1050°F and operating conditions subject to thermal or mechanical fatigue. The few failures which have occurred have generally been due to other weld defects such as lack of penetration.

Without backing rings, these pipe welds might have contained substantially more serious defects such as severe lack of penetration, concavity, burn through or cracking. Such defects have led to failures in piping systems where sound welds made with backing rings have not failed.

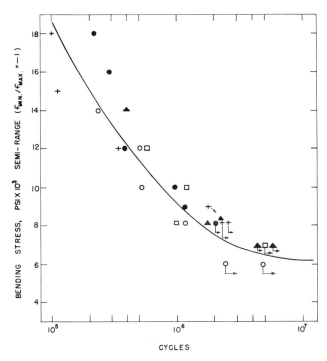

Fig. 12-5. Results of fatigue tests of 6⅝-in. diameter by ⅜-in. wall steel pipe butt welds with backing rings.[8] (The five types of plotted points indicate the number of sources of specimen supply.)

BURN THROUGH

Definition

Burn through refers to a coalescence of weld metal beyond the root of the weld. Where droplets form, the burn through is often described as *icicles* or *grapes*. Burn through may also involve a melting of metal away from the root of the weld, thus forming a cavity through a backing ring or strip.

Examples of icicles and cavity formation in pipe welds made with backing rings are illustrated in Figs. 12-6 and 12-7. Similar conditions of icicles may occur on the inside of pipe welds made without backing rings. Concavity in such welds is discussed under *Sink*, see Fig. 12-28.

224 *Defects and Failures in Pressure Vessels and Piping*

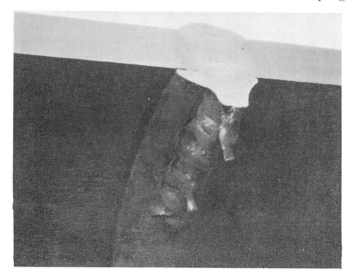

Fig. 12-6. Example of burn through involving icicles on underside pipe butt weld.

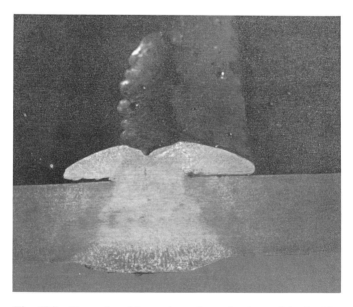

Fig. 12-7. Example of burn-through cavity formed in backing ring on underside of pipe weld.

Weld Defects 225

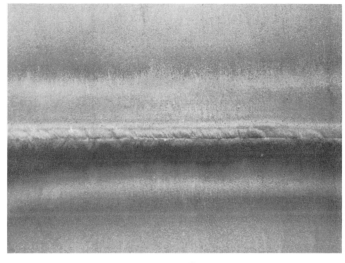

(A)

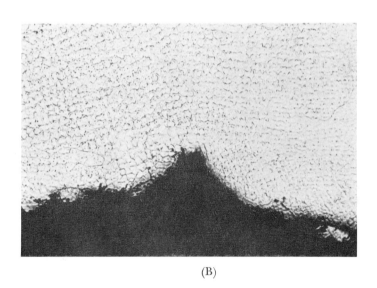

(B)

Fig. 12-8. Centerline crevice in root of butt weld in fully austenitic stainless steel pipe. (A) Photograph of inside pipe surface. (B) Cross section across center of weld. (Magnification 100×)

Service Experience

Unless very severe and continuous, burn through conditions generally are not considered harmful.

CENTERLINE CREVICE

Definition

A centerline crevice or crease represents a shallow linear groove formed by shrinkage or upsetting on the underside of a root bead.

An example of a crevice in a weld made by inert-gas tungsten-arc welding is shown in Fig. 12-8. The material is a highly alloyed stainless steel.

Service Experience

Although a centerline crevice may appear on a radiographic film similar to lack of penetration, it is not considered harmful or rejectable. In pipe joints involving austenitic materials, this condition may not be readily avoidable.

CRACKS

Definition

Cracks represent linear ruptures of metal under stress. Although sometimes large, cracks are often very narrow separations within the weld or adjacent base metal.

Typical major weld cracks are illustrated in Fig. 12-9.

Cracking in welded joints occurs in several forms, normally described as follows:

(1) Hot cracking in weld deposits
(2) Cold cracking in weld deposits
(3) Microfissuring
(4) Base-metal cracking

The subject of weld-metal cracking has been widely discussed in the literature. A literature survey published in 1952 covered 844 references.[10]

Distinction should be made between major cracks as may be readily

Weld Defects 227

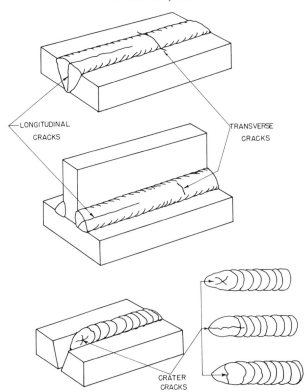

Fig. 12-9. Typical cracks in weld metal.[52]

visible and fissures or microfissures requiring special techniques for their detection, Figs. 12-10 and 12-11. Whereas major cracks generally are considered harmful, fissures or microfissures may not reduce the service life. The limiting size of a fissure (or crack) may vary with service conditions. The terms fissures and cracks often are used interchangeably. Thus, their separation by an arbitrary size factor is not possible. It would be just as difficult to predict when a fissure becomes a crack.

Since fissures are in essence separations present in every commercial metal, the term fissure is used here to describe very small cracks not normally considered critical to the service life, and not likely to propagate under normal operating conditions. Nevertheless, the mechanism of fissure formation as a result of welding (or heating), is referred to as cracking (either as hot cracking or as cold cracking).

Fig. 12-10. Visible surface hot crack in Type 347 stainless steel weld joining flange to 1½-in. diameter bellows section. Crack delineated by liquid penetrant.

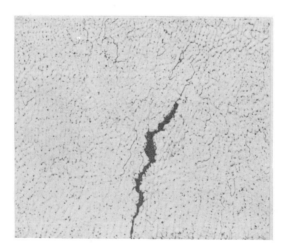

Fig. 12-11. Small fissuring in Type 347 stainless steel weld deposit. (Magnification 100×)

Hot Cracking

Hot cracking occurs at elevated temperatures during cooling after the weld metal has been deposited and started to solidify from the molten state. Normally, it is related to hot shortness. Since hot cracking is generally associated with grain boundary liquation or brittle grain boundary phases and inclusions, hot cracking appears generally as intergranular fissures or interdendritic * fissures.

As holds true generally for cracking conditions, the principal factors causing hot cracking are mechanical and metallurgical. The mechanical factors include shape and thickness of the material, joint design, and size and shape of the weld bead. As a general rule, the severity of the cracking increases with increases in the strains imposed on the weld during solidification. The principal metallurgical factors are the presence of segregations or liquid phases along the grain boundaries. The latter tend to remain liquid to lower temperatures than the weld metal does normally by itself.[11-13]

Ferritic Steels. The most harmful elements causing hot cracking in ferritic steel weld deposits are in decreasing order of potency: [14-17]

$$S, B, P, As, Cb, Sn, Sr, Ta, \text{ and } Cu$$

In steel, under conditions of restraint, the carbon content and manganese/sulfur ratios of the resulting weld deposit have a significant bearing on hot cracking. Tests [11, 18-20] have shown that up to about 0.112% carbon, the manganese/sulfur ratio is a major factor. Above this value, the carbon content becomes increasingly significant. Above 0.15% carbon, the susceptibility to cracking tends to be quite marked under severe restraint, even with high manganese sulfur ratios. While at about 0.12% carbon, a manganese/sulfur ratio exceeding 20:1 is generally sufficient to prevent hot cracking, at 0.15% carbon, the manganese/sulfur ratio should exceed 55:1. In steels with relatively low manganese/sulfur ratios, the ductility is lowered in the 1750 to 1920°F temperature range.[20] Low manganese/sulfur ratios are generally avoided in commercially produced pressure vessel and pipe steels. Thus, hot cracking is not a common experience and need not be a concern in normal fabrication.

---

* Dendrites describe the tree-like branching patterns in the grain structure in castings and weld deposits, see Figs. 8-12 and 12-11.

Silicon and nickel may also increase the susceptibility to hot cracking.[16] The effects of these elements may be interrelated with sulfur. Thus, silicon and nickel may increase the segregation of sulfur along the grain boundaries.[16, 21]

The effect of some residual elements in the quantities normally encountered in steel may be inconsequential. For example, arsenic in an amount less than 0.10% has not been found to produce cracking.[22] This 0.10% limit, which may even be conservative, is two to three times higher than the amount present in normal steels.

Austenitic Stainless Steels. Hot cracking in austenitic stainless steel welds has been extensively discussed in the literature.[23-25] The following elements are most generally responsible for hot cracking in decreasing order of potency:[11, 13, 26, 27]

$$B, S, P, Cb, As, Sn, Pb, Zr, Ta, \text{ and } Cu$$

Because of many chemical and metallurgical interrelations, any "potency" listing is very generalized. For example, boron when present in minute amounts of 0.001 or 0.01% very greatly enhances crack formation. If, however, the boron content is increased to ½ to 2%, hot cracking is minimized and high temperature strength is greatly increased.[28] The tendency for hot cracking is reduced by the formation of a complex Fe-Ni-Cr-B eutectic dissolving Si, P and other elements which cause liquation.[28]

The effects of carbon and nitrogen depend also on composition. They tend to reduce the cracking tendency of some elements such as Cb or Ta by forming carbides, or contribute to cracking of other elements by reducing the presence of ferrite. Other elements further affect the cracking tendency. Mn generally reduces the cracking tendency in weld deposits. Another beneficial effect of Mn is that it tends to combine with sulfur. However, when copper is present, sulfur combines with Ni and Cu to form a low melting Ni-Mn-Cu sulfide which produces hot shortness. Si enhances the cracking tendency in the presence of P. Ni reduces the cracking tendency of Cu. Ta adds to the crack susceptibility of Cb. Since many trace elements are present in base and weld metals, their interrelations are often difficult to evaluate.

Examples of major and minor hot cracks in Type 347 stainless steel weld metal are shown in Figs. 12-10 and 12-11.

Weld Defects

To provide maximum resistance to hot cracking in austenitic stainless steel weld deposits, it is normally considered desirable that the weld deposit contain a two-phase structure where the second phase involves approximately 3 to 8% ferrite. This is accomplished primarily by careful control over filler metal composition by maintaining the ferrite-forming elements such as chromium and molybdenum on the high side of the specified range, and the austenite-forming elements, such as nickel at the low end. The Schaeffler diagram,[29] Fig. 12-12, is widely used as a convenient method to estimate the amount of ferrite in weld metal. It is based on chemical composition. Type 308 and 347 weld metals normally contain a small amount of ferrite to avoid hot cracking. This is particularly important in Type

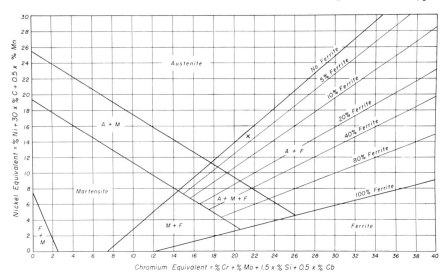

Fig. 12-12. Schaeffler diagram for estimating structure in Cr-Ni stainless steel weld deposits.[29]

Example: Point X on the diagram indicates the equivalent composition of a Type 318 (316 Cb) weld deposit containing 0.07% carbon, 1.55% manganese, 0.57% silicon, 18.02% chromium, 11.87% nickel, 2.16% molybdenum and 0.80% columbium. Each of these percentages was multiplied by the "potency factor" indicated for the element in question along the axes of the diagram, in order to determine the chromium equivalent and the nickel equivalent. When these were plotted, as point X, the constitution of the weld was indicated as austenite plus from 0 to 5% ferrite; magnetic analysis of the actual sample revealed an average ferrite content of 2%.

For austenite-plus-ferrite structures, the diagram predicts the percentage ferrite within 4% for the following stainless steels: 308, 309, 309 Cb, 310, 312, 316, 317 and 318 (316 Cb).

347 and other columbium-bearing stainless steels, which are highly susceptible to hot cracking.[26, 30]

Because of the high nickel content, Type 310 (25Cr–20Ni) normally exhibits a fully austenitic structure which is more susceptible to hot cracking. Type 312 (29Cr–9Ni) usually contains a large ferrite content. Thus, this grade is particularly suitable for welding stainless steel base metals high in nickel, manganese, and carbon.

Hot cracking is by no means limited to ferrous alloys.

Aluminum Alloys. In aluminum alloys, hot cracking tends to occur in the following alloys: [31]

Binary alloys	Si 0.0 — 1.5%
	Cu 0.5 — 5.0%
	Mg 1.0 — 4.0%
Ternary Alloys	Cu 0.0 — 1.0%, Si 0.5 — 1.5%
	Mg 0.5 — 2.5%, Fe + Si 0.5 — 1.0%
	Mg 0.5 — 2.0%, Cu 0.0 — 2.5%
	Si 0.1 — 0.5%, Fe 0.25 — 0.5%
	Mg 0.5 — 2.7%, Mn 0.0 — 0.1%

An example of hot cracking was illustrated in Fig. 7-28.

The welding filler metal after dilution * with the base should not contain an excess of embrittling phases such as Mg_2Si, $CuAl_2$, $AlFeSi$, $MnAl_6$, etc.[31]

The crack susceptibility also varies significantly with the welding process and procedure. For example, Al–5Cu–0.1Cd alloys exhibited extensive weld cracking in welds made by oxyacetylene welding. Cracking was absent in welds made by inert-gas tungsten-arc welding.[32]

Copper Alloys. In copper alloys, as little as 0.005 Pb and 0.0006 Bi can cause hot cracking in weld deposits.[33] The cracking tendency is reduced in alloys containing at least two structural components. For example, aluminum bronze filler metals with about 7.5% Al are highly susceptible to hot cracking.[34, 35] Filler metals with about 10

---

* Dilution refers to the mixing of weld metal with that part of the base metal which is molten by the welding operation. Depending on the welding process, procedure and joint configuration and other variables, dilution may vary between 10 and 90%.

Weld Defects

to 12% aluminum, resulting in a two-phase structure, show little if any susceptibility to hot cracking unless dilution again changes the composition.

Copper-nickel alloys are subject to hot cracking in the presence of the following elements:

Pb, Bi, S, P, Cd, Sb, Se, Te, and C

Nickel Alloys. In nickel alloy welding, Pb, Bi, S, P, Zr, and B promote hot shortness and cracking.[36] Nickel is the most susceptible, whereas "Inconel" is more resistant to hot cracking.[37]

In small amounts, Al, Ti, C, Mo, and Si generally are considered either beneficial or innocuous.[36] However, these elements may induce hot cracking when exceeding specific maximum limits. This may explain, for example, why C, Si, Ti, Cu, O, and N have been reported to promote hot cracking in "Inconel." [38]

Effect of Welding Process

Since conditions for hot shortness are usually more prevalent in base metals, hot cracking may occur more readily in those welding processes involving substantial melting of the base metal; i.e., mixing of base metal and weld metal. This can occur with submerged-arc welding and deep penetration coated electrode welding.

Cold Cracking

Cold cracking in steels refers to cracking which occurs below 400°F, usually near or at room temperature. It is sometimes *delayed*, occurring hours or even days after the weldment has equalized in temperature and is, therefore, free of thermal gradients and thermal stresses.[39]

In general, cold cracking starts in the heat-affected zone unless the weld metal exhibits a higher hardenability than the base metal.[39] Even within the same material, cold cracks do not grow uniformly in length and depth. While some cracks do not grow for a long time, and then suddenly grow rapidly, others grow rapidly from the start. Figure 12-13 illustrates an example of crack growth rates measured in an alloy steel weld approximately $5/8$ in. thick.[40] Cracking occurred in the heat-affected zone of a single-pass submerged-arc butt weld.

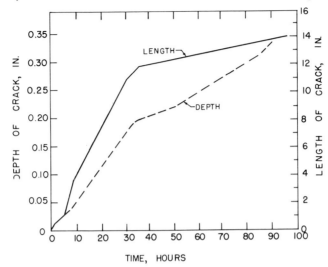

Fig. 12-13. Example of the rate of cracking in a ⅝-in. thick alloy steel plate containing approximately 0.35 C, 3 Cr, 3 Ni and 1 Mo.[40]

In steels, cold cracking is primarily associated with the combined effects of hydrogen, restraint, and martensite formation. Increasing the carbon content in the base metal and the manganese content in the weld also tends to promote cold cracking.[18] The effect of hydrogen in causing or contributing to cold cracking has been widely discussed in the literature.[39, 41-47] To minimize the tendency of hydrogen to cause cold cracking in ferritic steels, particularly alloy steels, electrodes with special so-called low hydrogen coatings are extensively used for welding pressure vessels and piping.

Microfissuring

Microfissures or *microcracks* are very small fissures not detectable at magnifications less than ten diameters. Often, they are visible at a magnification of 100 diameters or higher.

Microfissures may be caused by hot or cold cracking. As hot cracks, they are generally intergranular. As cold cracks, they are predominantly transgranular, though they may also be intergranular.

Intergranular microfissuring (hot cracking) which occurred at elevated temperatures may propagate in a transgranular path due to cold cracking at room temperature, particularly in the higher strength alloy steels.[48]

In extremely small sizes, microfissures are probably present in the majority of welds as well as in base metals.

Base-metal Cracking

Welding also causes cracking in the adjacent base metal, usually in the heat-affected zone.

Depending on the material being welded, either *hot or cold cracking*, or both, may occur. Sometimes, the cracks may start in the base metal and continue into or through the weld deposit.

Hot cracking in the base metal caused by welding generally is the result of the same factors responsible for hot cracking in weld deposits. The occurrence of base-metal cracking in the heat-affected zone of welds in columbium-bearing stainless steel after service at elevated temperatures of over 1050°F was reviewed in Chapter 4.

Such base-metal hot cracking occurs also frequently immediately after welding. An example in Type 347 stainless steel is illustrated in Fig. 12-14. This type of hot cracking, usually extending for only a few grains, is probably the result of the welding stresses rupturing the grain boundaries in the hot-short base metal. This type of cracking can be minimized by reducing welding stresses as, for example, by depositing weld metal of high ductility (see p. 289, Chapter 13).

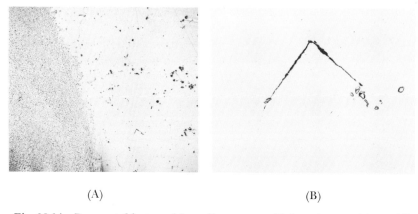

(A)   (B)

Fig. 12-14. Base-metal hot cracking adjacent to weld deposit caused by welding in Type 347 stainless steel. (A) Etched—(Magnification 50×), (B) Unetched—(Magnification 500×)

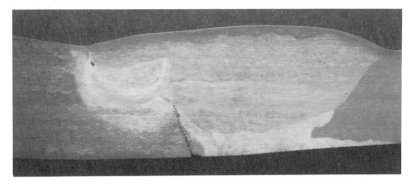

Fig. 12-15. Example of toe cracking in a 5 Cr–½ Mo alloy steel casting.

Where cracking starts immediately adjacent to the fusion boundary of the weld, the fissures may have been initiated by melting (liquation) along the grain boundaries.[49]

Although the columbium-bearing stainless steels are generally considered the most susceptible of the austenitic stainless steels to base-metal hot cracking, the other stainless steel grades such as 321, 316, and 304 have also been found susceptible.[50, 51]

Base-metal *cold cracks* often appear as *underbead cracks* occurring essentially parallel to the weld, Fig. 14-2. However, they may also occur as *toe cracks* starting at the toe of a weld, Fig. 12-15.

The tendency toward cold cracking varies with different welding conditions such as welding process, procedure, filler metal composition, restraint, weld length, preheat, postheat, etc.

Although specific welding variables make predictions difficult, the tendency toward cold cracking is influenced by the following factors:[52, 53]

Factors reducing cracking tendency:

Preheat
Postheat immediately after welding
Increasing arc energy
Skip welding

Factors increasing cracking tendency:

Increasing wall thickness
Increasing restraint

Weld Defects

Increasing strength of base or filler metal
Increasing hardenability of base or filler metal
Increasing length of weld

Welding process and procedure variables are also important. For example, on hardenable steels, welds made with the gas-shielded consumable metal-arc process tended to develop a greater tendency to cracking than those made with the submerged-arc process. The crack susceptibility was reduced further in welds made with the inert-gas tungsten-arc process.[53]

Test Results

In one set of tension tests, the length of the internal weld cracks in mild steel welds and in the presence of high residual stress levels was not dependent on the amount of static stress required for fracture initiation of cracks $1\frac{1}{2}$ to 3 in. in length, the maximum length investigated.[54]

In fatigue tests, specimens $2\frac{3}{8}$ in. wide by $\frac{1}{2}$ in. thick containing fine internal cracks parallel to the weld direction and amounting to approximately 5% of the total cross-sectional area reduced the fatigue strength in 10,000,000 cycles by approximately 60%.[55]

Surface cracks, as may occur in the root weld in pipe joints, are extremely critical in fatigue tests (and service) and will produce severe reduction in fatigue strength.[56]

Service Experience

Microfissuring. It is generally recognized that fully austenitic stainless steel weld deposits are susceptible to microfissuring during cooling soon after solidification (hot cracking). This has been observed in 1% columbium-stabilized 18Cr–12Ni–1½Mo weld deposits producing fully austenitic weld metal. This composition has been used extensively in high-temperature steam power plant service in Germany. Service history involving ten years indicates that the internal microfissures present have not grown and propagated into failures.[57-59]

Weld-Metal Cracking. Major cracking is considered to be critical and has led to frequent service failures. In weld deposits, the first weld bead deposited is generally most susceptible to cracking. If the first deposit is crack-free, then it is likely that the balance of the weld

238   *Defects and Failures in Pressure Vessels and Piping*

deposit normally will also be free of cracks. However, this does depend on welding conditions, see Fig. 13-1. Exceptions may involve also severe base-metal defects such as laminations which may start cracking in the weld (see Fig. 13-25).

Base-metal Cracking. Base-metal cracking such as toe cracking and severe underbead cracking is also considered very critical and has resulted in service failures.

CRATER PITS

Definition

A crater pit represents an approximately circular surface condition extending into the weld metal in an irregular manner. It is caused by volumetric contraction of molten metal during solidification, usually the result of abrupt interruption of the welding arc.

An example of a crater pit in the weld root on the inside of a pipe is shown in Fig. 12-16. Unless cracking starts from these pits, they are generally not considered a cause for rejection. Occasionally, however, cracking may start from craters. Sometimes, they are also associated with incomplete fusion, Fig. 13-4. Then defects may be harmful, as discussed in the appropriate sections.

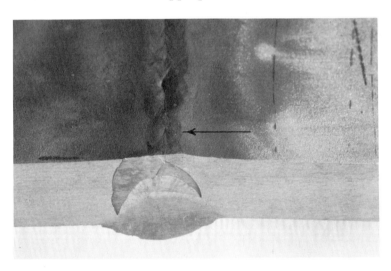

Fig. 12-16. Surface appearance of pipe inside of crater pit in root of stainless steel pipe butt weld.

Test Results

Cantilever-beam type fatigue tests on 2-in. and 4-in. Schedule 40 pipe welds made by inert-gas tungsten-arc welding indicated that crater pits did not cause or contribute to fatigue cracking.[9]

Service Experiences

Unless cracks or incomplete fusion (Fig. 13-4) were present in the original weld, crater pits have not been responsible as primary causes acting as crack starters in welds which failed in service.

HI-LOW

Definition

Hi-Low represents mismatch or misalignment of the plate or pipe ends across the weld root. It may be caused by ovality of the pipe or pressure vessel ends, by differences of inside diameters, by problems of fit-up, by nonuniform weld end preparations, or by other causes.

Where joints are not accessible from the inside, mismatch may be minimized by special internal machining. Except for heavy wall thicknesses, internal machining is not always possible or practical.

In Fig. 12-17 typical examples of mismatch which occur in pipe butt joints are illustrated.

Test Results

Fatigue tests usually indicate a loss in fatigue life due to mismatch. For example, tests involving $1/8$-in. thick aluminum specimens with misalignment up to 0.020 displacement of the axes showed that under axial loading the misalignment lowered fatigue life severely.[60] However, in bending fatigue, this misalignment was of little consequence, Fig. 12-18.

Service Experience

Where the weld joints have exhibited proper penetration and fusion in the weld root, service failures due to mismatch have been difficult to prove. Exceptions have been environments involving severe mechanical or thermal fatigue or stress corrosion.

240 *Defects and Failures in Pressure Vessels and Piping*

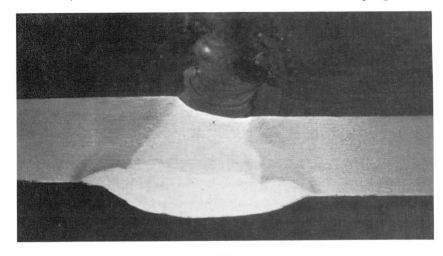

(A)

(B)

Fig. 12-17. Cross sections of pipe welds illustrating typical conditions of mismatch. (A) Concave. (B) Convex.

Weld Defects

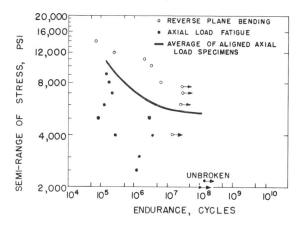

Fig. 12-18. Effects of misalignment on fatigue life of 1/8-in. thick aluminum specimens welded from one side and displaced by up to 0.020 in.[60]

In general, the consequences of mismatch should not differ significantly from those caused by sink (concavity).

INCOMPLETE FUSION

Definition

Incomplete fusion involves lack of complete melting and coalescence (fusion) of some portion of the metal in a weld joint. It may occur either between weld beads or between weld and base metal.

An example of incomplete fusion between weld beads is shown in Fig. 13-4. Incomplete fusion between weld and base metal is illustrated in Fig. 13-8.

Test Results

Some fatigue tests on pipe welds have indicated that incomplete fusion within the weld was of relatively little consequence.[61] The pipe specimens containing extensive lack of fusion were tested by alternate bending, with the weld located normal to the direction of the applied stress. The plane of the defect was circumferential and normal to the direction of stress. However, through the weld thickness, the defect followed the original 40 degree pipe edge bevel preparation from the outside surface to approximately the mid-thickness of the pipe wall. In the bending fatigue test, cracking was consistently initiated at the pipe inside at the root edge of the weld between pipe and backing ring. In only one instance was a minor localized fatigue crack initiated from the tip of the defect.

Service Experience

Only a few instances of service failures due to incomplete fusion on the inside of welds have been reported. Where failures did occur, the lack of internal fusion normally was quite severe, amounting to over 10 to 30% of the effective wall thickness.

Incomplete fusion at the inside or outside surface has been of greater consequence. Failures have resulted due to mechanical or thermal fatigue or stress-corrosion cracking.

LACK OF PENETRATION

Definition

Lack of penetration involves incomplete penetration of the weld through the thickness of the joint. It usually applies to the initial weld pass or passes made from one or both sides of a joint.

In double-welded joints, lack of penetration may occur within the wall thickness as a "buried" defect, Fig. 13-20. In pipe butt welds made from one side, lack of penetration may be below the weld root representing a surface defect, Fig. 11-7. At this location, it represents a more severe defect and acts as a surface notch.

Test Results

The static strength tolerance for lack of penetration is low, particularly when the weld reinforcement has been removed. This is illustrated in Fig. 12-19 showing the effects of lack of penetration in double-welded aluminum plate. Somewhat higher strength values were obtained on specimens retaining the weld reinforcement.[62]

Orientation of lack of penetration can also be important. Tests on plate specimens have shown [55, 63] that if the fault is oriented in the same direction as the load, a considerable degree of lack of penetration will not adversely affect the fatigue strength. The same degree of lack of penetration, where the defect is oriented transversely to the applied load, results in a severe reduction of fatigue strength.

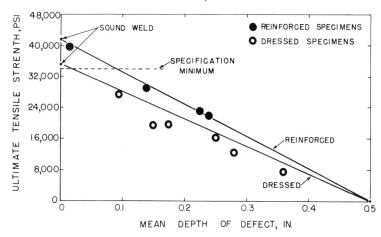

Fig. 12-19. Effect of depth of lack of penetration on tensile strength in double welded aluminum pipe joints ½-in. thick.[62]

In ⅞-in. thick plate, fatigue test specimens with internal lack of penetration of 40% severity reduced fatigue strength by 64%. At 60% severity, the fatigue strength was reduced by 69%. The loss in strength was even greater when the weld reinforcement was removed,[63] so that weld reinforcement was considered beneficial.

Tension fatigue tests of double-welded aluminum plate conducted on ¼-in. and ½-in. thick material showed that a slight lack of penetration was not harmful, provided that the defect area is less than 7% of the cross-sectional area of the plate.[62]

A 1-in. long lack of penetration defect in the center of a ¾-in. thick by 4-in. low-alloy steel (HY-80) fatigue test specimen failed after 41,600 cycles as compared to an average life of 150,000 cycles for sound butt welds.[64]

Thermal fatigue tests on 8-in. diameter by 0.906-in. thick pipe containing lack of penetration in the center of the pipe 2 in. long by 3/16 in. deep by 0.027 in. wide as may occur in circumferential butt welds has neither caused nor contributed to cracking in tests involving heating to 1000°F followed by water quenching to below 200°F. Typical thermal fatigue cracks were initiated after about 250 cycles on the inside pipe surface involving sound metal. The cracks have been propagated gradually into the pipe wall and show no relation to the center lack of penetration, Fig. 12-20.

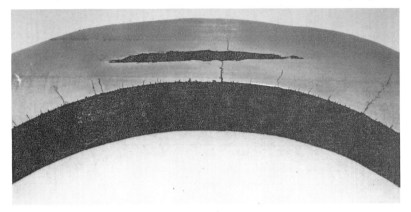

Fig. 12-20. Cross section through milled 3/16-in. notch in center of pipe wall in 0.906-in. thick pipe which did not cause or contribute to cracking in thermal fatigue test.

Mechanical fatigue tests on pipe butt welds with lack of penetration on the inside surface have shown that this resulted in a reduction of fatigue strength of over 50%.[61]

On 3/8-in. thick pipe specimens subjected to bending fatigue testing, 1/8-in. deep continuous lack of penetration at the root of butt welds reduced fatigue strength measured as a 60% reduction of stress in 2,000,000 cycles.[61]

Service Experience

Lack of penetration is one of the most critical defects and has resulted in service failures in pressure vessel, tank, and pipe welds,

Fig. 12-21. Lack of penetration in 20-in. diameter by 0.375-in. wall weld in gas distribution service which has not caused failure after 35 years of service. (Magnification 2×)

Weld Defects 245

see Chapter 13. When it occurs at the inside surface of a pipe weld, failures may result in service involving mechanical or thermal fatigue or stress corrosion.

However, in noncritical service applications, lack of penetration may not result in cracking and service failures. Figure 12-21 illustrates the cross section of a butt weld made by oxyacetylene welding in 1928. Failure of the 20-in. diameter by 0.375-in. wall gas-distribution piping has not occurred after 35 years of service.

OVERLAP

Definition

Overlap represents an excess (overflow) of weld metal which extends beyond the limits of fusion over the surface of the base metal. An example of overlap on a butt weld is shown in Fig. 12-22.

Overlap is more often associated with fillet welds than with butt welds. It results in an apparent increase in the size of the weld, which may lead to erroneous estimates of the size (and strength) of fillet welds.[65]

Service Experience

In butt welds, minor overlap has not resulted in service failures.

Severe overlap in fillet welds has led to failure because of insufficient joint strength.

OXIDATION

Definition

Surface oxidation is the result of insufficient protection from the atmosphere of the weld and adjacent base-metal surface. Surface oxidation is of concern generally only on the underside of weld joints made without backing rings.

Surface oxidation may range from slight discoloration to severe oxidation on stainless steels. This severe oxidation also referred to as "cauliflowering" is illustrated in Fig. 13-13.

Less severe oxidation may occur on the underside of welds made of carbon, low-alloy steels and nickel alloys, where the surface appears slightly wrinkled. This condition, also described as "sugaring," is illustrated in Fig. 12-23.

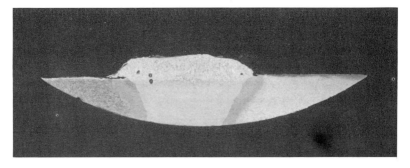

(A)

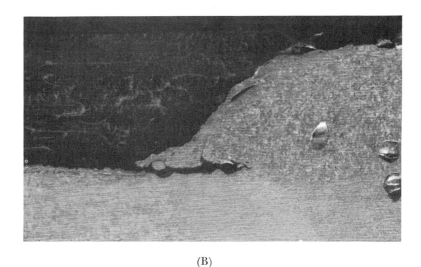

(B)

Fig. 12-22. Example of overlap on pipe butt weld. [Magnifications (A) 1×, (B) 6×]

Test Results

Cantilever-beam type fatigue tests on 2-in. and 4-in. diameter steel and cupro-nickel pipe butt welds have indicated that sugaring does not affect fatigue life.[9]

Service Experience

There has been no evidence that surface oxidation of the type illustrated in Fig. 12-23 has resulted in cracking or service failures.

Weld Defects 247

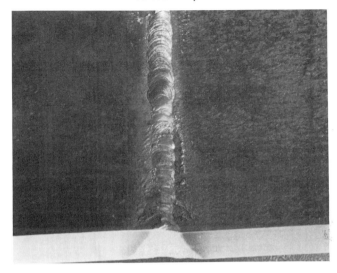

Fig. 12-23. Surface oxidation resulting in sugaring on root surface on inside of weld 1¼ Cr–½ Mo alloy steel pipe welded by inert-gas tungsten-arc process.

POROSITY

Definition

Porosity [45, 66, 67] is the presence of gas pockets or voids (usually spherical in shape) caused by the entrapment of gas evolved during weld metal solidification. Spherical gas pockets were previously illustrated in Fig. 5-2. Nonspherical voids entrapped along the grain boundaries in a Type 347 weld deposit are shown in Fig. 12-24.

Sometimes elongated tubular gas pockets are also described as *worm holes* or *piping*. Large isolated gas pockets may be referred to as blowholes. These may be partially filled with slag.[65]

Test Results

Static tension, bend and impact tests on butt welds in quenched and tempered T-1 low-alloy steel have shown that porosity in quanti-

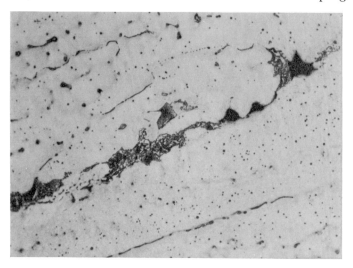

Fig. 12-24. Voids entrapped between grains (dendrites) in a Type 347 stainless steel weld deposit. (Magnification 500×)

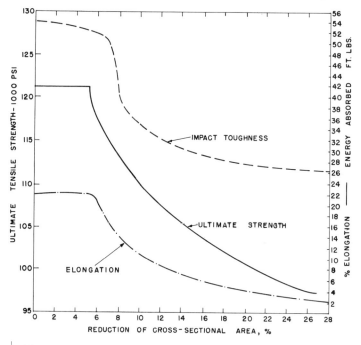

Fig. 12-25. Effects of porosity on ultimate strength, elongation and Charpy V-notch impact toughness of submerged arc welds.[68]

ties reducing the cross section up to 5% have no significant effect on strength, ductility and toughness.[68] The test results are summarized in Fig. 12-25. However, when located near the surface, even small porosity could lead to failure in bend tests.

In static tension tests, porosity appears to affect fillet welds far more adversely than butt welds. For example, in one Al-Mg-Si-Mn alloy, porosity reduced the tensile strength by 30 to 60%.[69]

In severe fatigue tests on welds in steel where the tests are continued until failure, such tests have shown that porosity levels above 3, 4, or 5% may reduce fatigue strength significantly.[70-73] In fact, a defect severity of 0.8% even reduced the mean endurance limit from 34,600 to 28,000 psi. The results of fatigue tests on mild and low-alloy steel specimens containing transverse welds with varying levels of porosity are shown in Fig. 12-26.[64, 71, 72, 74] The three curves in

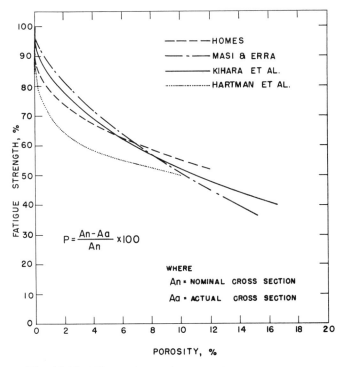

Fig. 12-26. Effects of porosity on fatigue strength of flat steel test specimens. (After Homes,[71] Masi and Erra,[72] Kihara et al.[74] and Hartman et al.[64])

250 *Defects and Failures in Pressure Vessels and Piping*

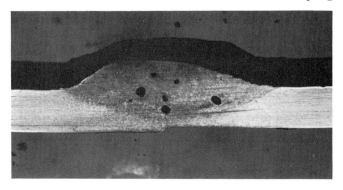

Fig. 12-27. Severe porosity which did not cause failure in an oxyacetylene weld after 40 years of service in gas distribution piping system. (Magnification 2×)

close agreement represent mild steel specimens. The data showing a somewhat more severe initial loss of fatigue strength were obtained on HY-80 low-alloy steel.

In 2-in. and 4-in. Schedule 40 steel and cupro-nickel pipe specimens subjected to cantilever-beam fatigue testing, aligned porosity involving 0.045-in. diameter holes spaced 1T apart (0.154 and 0.237 in.) and located $\frac{1}{16}$ in. above the root did not reduce fatigue life.[9]

Service Experience

An illustration of severe porosity is shown in Fig. 12-27 in a weld made by oxyacetylene welding. Even after nearly 40 years of service in a gas distribution piping system, the porosity present has not resulted in, or contributed to, a service failure.

Very rarely has porosity been associated with actual service failures. No failures have been reported in welds where the porosity was within applicable Code requirements. Quite likely, porosity limits two to four times those now established in Codes could be tolerated in the majority of critical service applications.

SINK OR CONCAVITY

Definition

Sink or suck-up in a root weld bead refers to concavity. It is produced by gravity sink of the molten metal or by surface tension of

Weld Defects 251

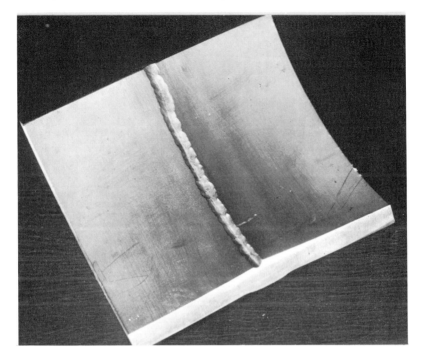

Fig. 12-28. Example of sink (concavity) in pipe weld not normally considered harmful.

the weld bevel pulling the molten weld metal into the bevel. Thus, concavity can be produced at the top (12:00 o'clock) position of a butt weld in a horizontal pipe section.

An example of sink in a pipe weld made by the inert-gas tungsten-arc welding process is illustrated in Fig. 12-28.

The occurrence of sink depends on various factors including weld joint preparation, welding procedure variables and materials. Too much heat in the second weld pass may also cause some suck-back in the first root pass weld, Fig. 12-29, particularly when the first root pass involves a thin cross section.

Severe sink can cause root pass cracking, Fig. 12-30, which may not be apparent by subsequent radiographic examination.

Although root concavity is most generally associated with welds made by the inert-gas tungsten-arc welding process, it can be caused also by other welding processes. Thus, pipe welds made without

252   *Defects and Failures in Pressure Vessels and Piping*

Fig. 12-29. Suck-back of part of weld root pass due to remelting caused by second weld pass.

backing rings by oxyacetylene welding or shielded metal-arc welding quite frequently contain some concavity.

Test Results

Cantilever-beam type fatigue tests on 2-in. and 4-in. diameter Schedule 40 cupro-nickel and steel pipe test sections have shown that root concavity $\frac{1}{16}$ and $\frac{1}{32}$ in. deep has reduced fatigue life to as low as 10 to 20% of the life of welds with uniform root penetration. Only $\frac{1}{64}$-in. deep root concavity was not found to reduce fatigue life.[9]

Service Experience

Excessive sink has been involved in a number of service failures involving severe mechanical or thermal fatigue. In several instances, cracking appeared to have occurred in the weld root where the welding was done originally.

The amount of concavity which can be tolerated will depend on the service requirements, material and joint dimensions.

In critical applications involving pipe wall thicknesses of over $\frac{1}{2}$ in.,

Weld Defects

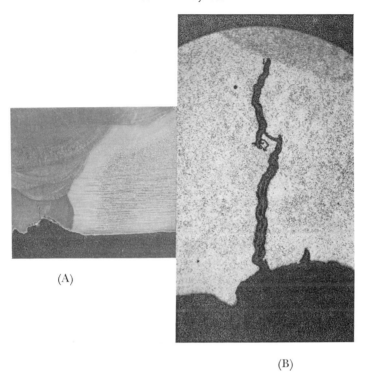

Fig. 12-30. Cracking in root pass due to excessive sink in 6-in. OD by 0.432-in. wall $2\frac{1}{4}$ Cr–1 Mo pipe butt weld. [Magnification (A) 4×, (B) 50×]

a $\frac{3}{32}$ in. or $\frac{1}{16}$ in. maximum limit or concavity will probably be similar to the changes in surface contour provided by $\frac{3}{32}$ in. or $\frac{1}{16}$ in. mismatch of pipe ends.

However, there are some extremely severe service applications where any kind of concavity may represent a hazard. Dissimilar metal joints in service involving severe thermal fatigue should be included in this category.

SLAG INCLUSIONS

Definition

Slag represents nonmetallic solid material entrapped in the weld deposit or between weld metal and base metal.

254 *Defects and Failures in Pressure Vessels and Piping*

Fig. 12-31. Cross section of weld joint containing slag inclusions after service for 20 years at 960°F, 1800 psi. (Magnification 7×)

Slag inclusions may appear as individual particles, Fig. 12-31, or as linear continuous or interrupted bands.

Test Results

Because of the many forms and shapes of slag inclusions possible and the difficulty in expressing slag on a quantitative basis, fatigue test results have been somewhat contradictory.[55, 75] Fatigue test results [55] on steel specimens at a stress range of ± 16,000 psi have shown the following:

Defect	Life Cycles
Faint intermittent slag lines	152,000; 526,000
Intermittent parallel slag line	121,000
Parallel slag lines	347,000
Heavy parallel slag lines	49,000

Pulsating flat tensile fatigue test specimens have indicated that in steel welds for similar defect sizes, slag is likely to reduce fatigue life by about 12% below the reduction produced by porosity, Fig. 12-26.[74] On the other hand, in low-alloy steel (HY-80) welds, the loss of fatigue strength by slag inclusion was essentially the same as the effect of porosity (Fig. 12-26).[64]

In 2-in. and 4-in. diameter Schedule 40 pipe test specimens subjected to cantilever beam-type fatigue, severe slag over $1\frac{3}{16}$ in. in length did not reduce fatigue life or strength. Slag within $\frac{1}{16}$ in. of the outside surface was of minor consequence. Slag stringers within $\frac{1}{16}$ in. of the inside pipe surface were also insignificant in welds made by inert-gas tungsten-arc welding with a consumable insert.[9] However, in welds made with backing rings, such slag did further reduce fatigue life.[9]

Service Experience

Slag inclusions inside a weld have rarely resulted in service failures, unless they occur along the surface, or were of a sufficiently large size to reduce significantly the strength across the wall thick-

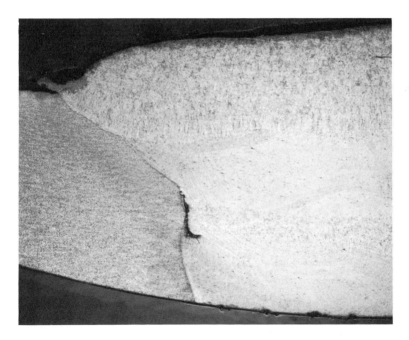

Fig. 12-32. Cross section through elongated liner slag inclusion after 25 years service at 825°F, 650 psi. (Magnification 5×)

256   *Defects and Failures in Pressure Vessels and Piping*

ness. The effect of slag in weld deposits should be considered similar to the effect of laminations and inclusions in steel plate or pipe. Figure 12-32 illustrates a slag inclusion in a weld in 12¾-in. diameter by 0.687-in. wall pipe joint which has not failed after 25 years of service at 825°F, 650 psi pressure.

SLUGGING

Definition

Slugging, also known as stubbing, refers to the addition of a separate piece or pieces of material in a joint before or during welding. It may be the result of entrapped welding filler wires or electrodes.

Slugging has been produced by placing welding electrodes into an open weld groove and depositing weld metal over them.

Instances of slugging occurred more frequently some 20 years ago when little radiographic inspection and spot-checking was done. Nevertheless, occasionally a slugged weld has been produced, even in recent years, and has failed subsequently. An example from a pressure vessel nozzle weld is shown in Fig. 12-33. Moreover, similar instances are still reported in the literature.[76]

Service Experience

Where strength welds are required, slugging can be very harmful.

TUNGSTEN INCLUSIONS

Definition

Metallic tungsten inclusions are particles which are deposited in the weld metal from a tungsten electrode used in the inert-gas tungsten arc welding process.

Examples of tungsten inclusions on the inside of a weld are shown in Fig. 12-34.

Test Results

Static tension tests and fatigue tests on aluminum and several aluminum alloys have shown that the presence of tungsten inclusions was of no consequence.[74, 77] The test specimens having a cross section of ¾ to ⅞ in. wide by 0.25 in. thick contained substantially larger

Weld Defects

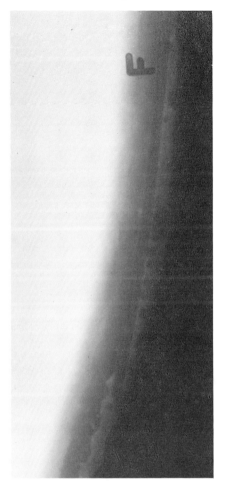

Fig. 12-33. Radiograph revealing slugging in nozzle weld of pressure vessel.

tungsten inclusions than those normally encountered, with some particles and clusters as large as $\frac{1}{4}$ in. Not a single fatigue crack leading to failure originated at a tungsten inclusion.

Service Experience

Tungsten inclusions generally are not considered harmful, unless their size and number become excessive.

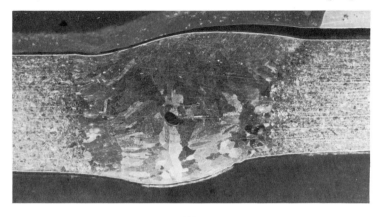

(A)

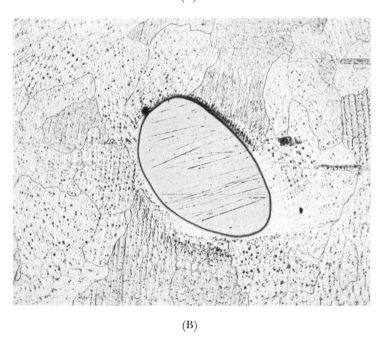

(B)

Fig. 12-34. Example of tungsten inclusion in Type 304 stainless steel weld deposit in 4-in. diameter by 0.237-in. wall pipe. [Magnifications (A) 5×, (B) 50×]

UNDERCUT

Definition

Undercut represents an intermittent or continuous groove or channel melted into the base metal adjacent to the toe or surface of a weld and left unfilled by weld metal.

Severe and slight undercut are illustrated in Fig. 12-35.

Test Results

Fatigue tests have illustrated that undercut can reduce significantly the fatigue strength. Typical pulsating tensile fatigue tests [74] on mild steel specimens have shown the following results:

Depth of Undercut	Pulsating Fatigue Strength at 2,000,000 Cycles (range of stress) (psi)
0.000	27,000
0.0236	19,900
0.0354	14,200

On low-alloy (HY-80) steel welds, fatigue tests on ¾-in. thick specimens have also shown adverse effects of undercut on fatigue life. Thus, undercut to a depth of 0.021 in. reduced fatigue life by about 10 to 20%, while undercut depth of about 0.050 in. reduced the fatigue life to about one-third.[64]

Root undercut on the inside of 2-in. and 4-in. pipe butt welds of depths of 0.01 to 0.02 in. and subjected to cantilever-beam-type fatigue has also reduced the fatigue life significantly.[9]

Service Experience

In service involving severe mechanical or thermal fatigue or pipe movement, undercut representing a sharp notch condition may result in cracking and failure.[65] An example is illustrated in Fig. 12-36.[78]

Applicable Codes covering critical applications normally do not permit undercut or limit its depth to 0.020 in.[79] Undercut is considered a notch condition and may thus have to be removed or reduced in depth by grinding. Actually, however, slight undercut is not uncommon, particularly on pipe and tank welds where the weld is made in the horizontal position. There is no evidence that such undercut has caused failures.

260 *Defects and Failures in Pressure Vessels and Piping*

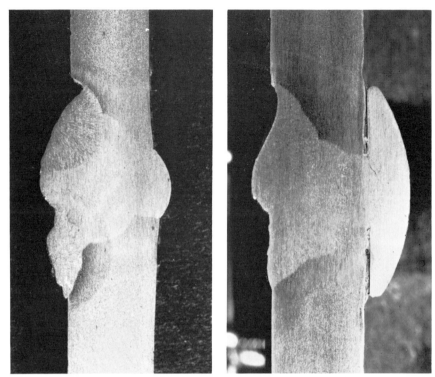

Fig. 12-35. Cross section of welds containing severe and slight undercut. (Magnifications 1½×)

Undercut should probably be considered in the same light as some surface laps and slivers on steel plate and pipe which, in fact, are permissible to a depth of 5%.

WAGON TRACKS

Definition

Wagon tracks are caused by voids or slag inclusions which are linear. They occur most commonly in the root of pipe welds between a backing ring which does not fit tightly and the edge of the weld. Fine indications of wagon tracks may also be caused by base-metal shrinkage adjacent to the weld edge between root pass and backing ring. In addition, wagon tracks can occur within the weld,

Weld Defects

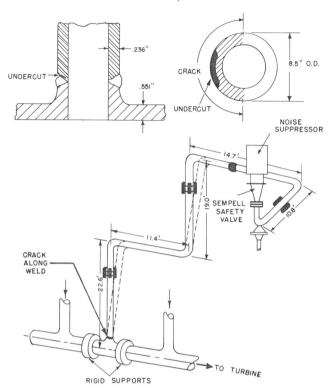

Fig. 12-36. Example of failure caused by severe undercut in service involving fatigue due to pipe movement.[78]

particularly between the first and second weld passes or layers, where slag remaining on the root stringer pass was not cleaned out of the crevice formed between the edge of the bead and the base metal.

On radiographic films, wagon tracks generally appear as parallel linear indications. A print of a radiograph and a cross section illustrating wagon tracks are shown in Fig. 12-37.

Service Experience

Wagon tracks in backing ring welds generally are not harmful, unless they involve actual incomplete fusion or lack of penetration within the weld root. Wagon tracks above the root pass which represent slag inclusions have the same effect as discussed earlier under slag inclusions.

262   *Defects and Failures in Pressure Vessels and Piping*

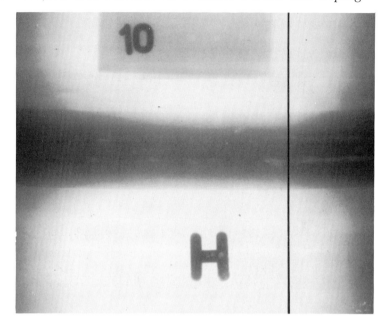

Fig. 12-37. Print of radiograph and cross section of pipe weld containing wagon tracks. (Magnification of cross section 1½×)

WELD REINFORCEMENT

Definition

The weld reinforcement represents the metal on the face of a weld in excess of the metal necessary for the specified weld size.

Whereas on pressure vessels the weld reinforcement is generally removed, it is frequently left intact on pipe joints. The weld reinforcement shape and appearance differs with the welding process and even the filler metal or electrode used.

Typical weld reinforcements of pipe welds are illustrated in Fig. 12-38 representing welds made by shielded metal-arc welding with E6010 and E7018 electrodes and by submerged-arc welding.

The *reinforcement shape* includes the height and width of the reinforcement, and the radius of the bead curvature at the junction between the reinforcement and the plate or pipe surface. The reinforcement angle is the primary parameter affecting fatigue strength. It is defined as the *angle subtended by the plate (or pipe) surface and a tangent to the weld reinforcement bead at its junction with the plate (or pipe)*, see Fig. 12-39.

Test Results

Fatigue tests have shown that the shape of the weld reinforcement can affect significantly the fatigue strength. This is generally related to the stress concentrations caused by the edge of the weld reinforcement.

Tension fatigue test specimens, in particular, have demonstrated that weld reinforcement will reduce fatigue life. Failures in these tests start generally at the edge of the weld reinforcement, at the top side of the weld. In these tests, fatigue strength of sound butt welds is generally increased by a significant amount by grinding or machining the reinforcement flush with the metal surface on both sides of the flat test specimen. However, the fatigue strength of welds containing severe defects is frequently not increased, and may be decreased by grinding off the reinforcement.[80]

Fatigue tests on steel welds [8] and on aluminum butt welds [81, 82] have shown that the more severe the reinforcement, i.e., when the reinforcement angle approaches 90 degrees, the lower will be the fatigue life. Typical fatigue tension test results are illustrated in

(A)

(B)

(C)

Fig. 12-38. Appearance of weld reinforcement of pipe welds made by shielded metal-arc welding with (A) E6010 and (B) E7018 electrodes and by (C) submerged arc welding.

Weld Defects 265

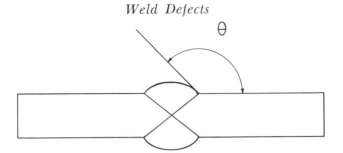

Fig. 12-39. Illustration of reinforcement angle.

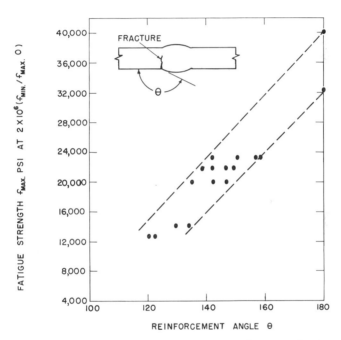

Fig. 12-40. Effect of weld reinforcement shape on the fatigue performance of steel butt welds.[8]

Fig. 12-40 for steel [8] and in Fig. 12-41 for aluminum.[82] The tests shown in Figs. 12-40 and 12-41 involved double-welded joints.

Similar results have been reported on steel test specimens.[83-85]

On 2-in. and 4-in. diameter Schedule 40 pipe, test specimens subjected to cantilever beam type fatigue, a $3/32$-in. high weld reinforcement did not reduce fatigue life.[9]

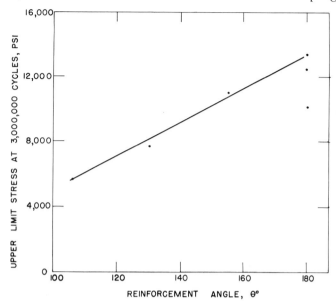

Fig. 12-41. Effect of weld reinforcement on fatigue performance of 3/8-in. thick Al-Mg-Mn alloy plates with reinforcement on both sides of test specimen.[82]

Service Experience

Service experience with welds in vessels and piping where the weld reinforcement has been left intact generally has not produced adverse results. Welds of the types illustrated in Fig. 12-38 have provided over 200,000 hours of satisfactory service in critical high-temperature high-pressure steam plant applications.

WELD SPATTER

Definition

Weld spatter involves particles of metal resulting from globules of metal unintentionally ejected from the molten pool during welding and deposited on the surface of the base or weld metal.

Service Experience

Weld spatter is generally not considered harmful.

WELD STRESSES*

A defect condition often not recognized involves residual welding stresses.

Figure 12-42 shows an ammonia tank 42-in. diameter by $\frac{3}{8}$-in. wall where an initial crack occurred across the seam weld after approximately a year and a half of service. The crack was repaired, but

Fig. 12-42. Side of 42-in.-diam carbon steel ammonia tank with welds $10\frac{1}{2}$ in. apart representing cracks which had propagated across $\frac{3}{8}$-in. vessel wall after one and a half to two years of service.

within a few months a second crack was apparent on the surface. This was repaired, but was followed by a third and later on, a fourth crack. Radiographs of the weld seam revealed two additional transverse cracks in the development stage, but not yet visually apparent since they had not reached the outside surface of the vessel (Fig. 12-43).

The seam weld had been made by automatic submerged-arc welding. It was not stress relieved, as this was not a requirement of the

---

*This is a part of a paper representing one of ten papers presented at the Second Conference on the Significance of Defects in Welds held in London May 29-30, 1968. The conference was organized by the Society of Non-Destructive Examination, the Non-Destructive Testing Society of Great Britain and the Welding Institute in behalf of the British National Committee for Non-Destructive Testing.

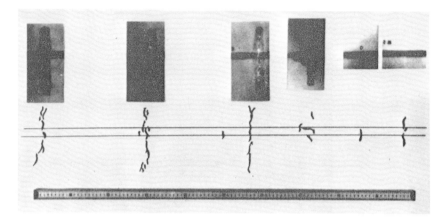

Fig. 12-43. Sketch of seam weld with prints of radiographic film to illustrate cracking and weld repairs in 42-in.-diam ammonia tank.

applicable codes. Nevertheless, cracking would not have occurred if the original weld had been stress relieved.

Welding stresses can also cause cracking in adjacent metals. Hot cracking in base metals, particularly those which are inherently hot short, is extensively discussed in the literature.

Even though a casting, forging or wrought pipe or plate material may not evidence cracking in the as-cast, forged or rolled condition, the material may nevertheless be inherently hot short at elevated temperatures and crack during welding. The welding stresses cause cracking in the heat-affected zone along the grain boundary of the base metal as a result of the low cohesive strength at elevated temperatures above 1400 F. An example is shown in Fig. 12-44.

In the higher carbon stainless steel castings such as HK-40[b], particularly when the grain size is course, the presence of significant grain boundary carbides may also cause cracking during welding. This cracking occurs near room temperature where the carbides are brittle and tend to rupture as a result of welding stresses.

Even where a material is not hot shot, cracking may occur as a result of severe welding stresses. As the weld cools and shrinks, the stress across the weld increases. Since the cohesive strength near room temperature generally is lower in the base metal than in the weld deposit, cracking may occur in the base metal. An example is shown in Fig. 12-45. Adjacent multiple weld repairs caused the

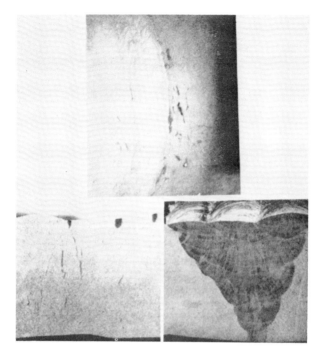

Fig. 12-44. Severe hot fissuring adjacent to weld in HK-30 5-in.-diam by $\frac{3}{4}$-in. wall stainless steel fitting.

Fig. 12-45. Cracking along weld edge caused by excessive welding without preheat. (Pipe diameter 15 in. by $1\frac{1}{4}$-in. wall.)

266d   Defects and Failures in Pressure Vessels and Piping

Fig. 12-47. Cracking in crotch of HK-30 stainless steel tees in methanol steam reformer furnace.

Fig. 12-49. Cross section through crack shown in Fig. 51 to illustrate extent and depth of shrinkage. ($7\times$— Reproduction 75% of original.)

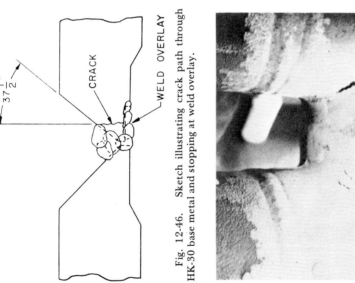

Fig. 12-46. Sketch illustrating crack path through HK-30 base metal and stopping at weld overlay.

Fig. 12-48. Closeup of typical cracks revealed by liquid penetrant examination.

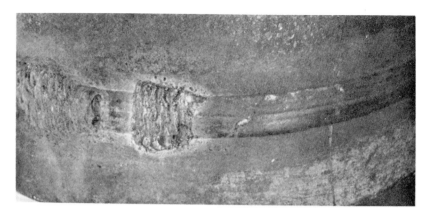

Fig. 12-50. Stress corrosion cracking in pipe weld made by inert-gas tungsten-arc root pass followed by submerged-arc welding. Radiograph of completed shop weld showed original weld to be sound. (Rough weld area represents repair weld made to seal cracks which occurred through original weld after several months of service.)

particularly severe stress, which led to the crack. The crack path is sketched in Fig. 12-46. The cracking did not penetrate into the weld overlay on the inside of the centrifugally cast stainless steel (HK-30[b]) tube.

Welding stresses may also open up already existing cracks or inclusions in castings.

Figure 12-47 illustrates cracking in the crotch area of a series of cast tees installed in a methanol furnace. A close-up is shown in Fig. 12-48. The castings are made of HK-30 alloy. At the ends, the outside diameter is $6\frac{1}{4}$ in. and the wall thickness is 0.63 in. In the crotch where the cracking is apparent, the wall thickness is 0.56 in.

Cross sectioning through a typical crack, shows that the cracking is the result of shrinkage which formed during the solidification of the casting, Fig. 12-49. In the as-cast condition, the cracks did not actually penetrate to the surface so that their presence was not detected by visual and liquid penetrant examination of the surface. However, after the circumferential butt welds had been made between the cast tees, the resulting normal residual welding stress was sufficient to propagate the cracking from the end of the shrinkage within the casting to the surface.

Even though in some service environments the presence of weld-

---

[b]HK-30 and HK-40 are 25 Cr-20 Ni stainless steel castings with approximately 0.35 per cent and 0.45 per cent carbon, respectively.

ing stresses may lead to cracking, the subsequent removal of the stress may stop further crack propagation.

Figure 12-50 illustrates an essentially "perfect" pipe weld made by inert-gas tungsten-arc root pass welding and completed by submerged-arc welding. The weld in the 14-in. diam. carbon steel pipe with a $\frac{3}{8}$-in. wall thickness was not stress relieved since, undermost United States Standards, such stress relief heat treatments are required in carbon steel pipe only for wall thickness of over $\frac{3}{4}$ in. The piping had been installed in a paper mill to carry process steam.

The paper mill maintenance welder made a very poor repair weld by shielded metal-arc welding into those areas in the pipe weld where the cracks had actually progressed through the weld wall thicknesses. On the inside of the pipe, severe weld craters were left in the repair weld areas (Fig. 12-51) as a result of burn-through. In other locations, the repair weld did not penetrate through the previously cracked weld, but covered up the cracks. However, this repair weld was subsequently stress relieved. No further cracking or crack propagation occurred during subsequent service (Fig. 12-52). The original cracking in the "perfect" weld was due to stress corrosion cracking. Elimination of the residual welding stresses, even though the repair weld was structurally very defective, eliminated the cracking problem. Thus, the repair-welded structurally defective weld, which provided satisfactory service, was actually more "perfect" than the original weld, which was structurally sound, but failed in service.

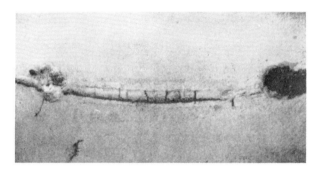

Fig. 12-51. Pipe inside after repair welding by shielded metal-arc welding process with weld craters apparent where welders "burned through" original weld.

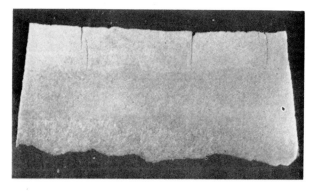

Fig. 12-52. Cross section showing cracks in original sound weld and confirming that cracks had not progressed into repair weld which had been stress relieved.

Fig. 12-53. Crack in 10-in.-diam pipe weld after 15 years of service.

Failures

A pipe weld which failed by cracking after approximately 15 years of service in a steam power plant is shown in Fig. 12-53. The cracking is detailed in Fig. 12-54. Radiographic inspection revealed the crack and also showed that the original weld was free of weld

266h   *Defects and Failures in Pressure Vessels and Piping*

defects (Fig. 12-55). This was confirmed by a review of the original radiographic films, on which the weld appeared structurally sound.

Thus, even welds without apparent flaws ("zero-defect welds") fail in service. Many examples of this could be cited.

Fig. 12-54. Detailed view of cracking shown in Fig. 12-53.

Fig. 12-55. Radiograph showing that the original weld was free of defects.

REFERENCES

1. Hargreaves, F., "Welding in Locomotive Construction and Repair," *Trans. Inst. Welding*, **12**, 98–104 (1949).
2. Winterton, K. and Corbett, J. M., "Dangers of Arc Strikes and a Possible Remedy," *Welding J.*, **39**, Res. Suppl., 121s (1960).
3. Ruge, J. and Woesle, H., "Results of Fatigue Investigations," *Die Schweisstechnik im Zeichen Neuzeitlicher Verfahren und Werkstoffe*, Deutscher Verband für Schweisstechnik, **23**, 59–68 (1962).
4. Kihara, H., Masubuchi, K., and Ogura, N., in "Researches on Weldable High Strength Steels" by Kihara, H., Suzuki, H., and Tamura, H., Soc. of Naval Architects, Tokyo, Japan (1957).
5. Yoshida, T. and Onove, in "Researches on Weldable High Strength Steels" by Kihara, H., Suzuki, H., and Tamura, H., Soc. of Naval Architects of Japan, Tokyo, Japan (1957).
6. Hauttmann, H., "Bursting Tests on Tubes Welded From Structural Steels Resistant to Cleavage Fracture," *Welding Research Abroad*, **4**, 32–40 (Oct. 1958).
7. Verein. d. Grosskesselbesitzer, 1961/62 Review.
8. Newman, R. P., "Fatigue Strength of Butt Welds in Mild Steel," *British Welding J.*, **7**, 169–178 (1960).
9. Meister, R. P., Beall, L. G., Randall, M. D., Hyler, W. S., and Martin, D. C., "Effects of Weld-Joint Imperfections on Piping Performance," Battelle Memorial Institute, Columbus, Ohio (July 31, 1964).
10. Williams, A. J., Rieppel, P. J., and Voldrich, C. B., "Literature Survey on Weld-Metal Cracking," Wright-Air Development Center, WADC Technical Report 52-143 (Aug. 1952).
11. Borland, J. C., "Suggested Explanation of Hot Cracking in Mild and Low Alloy Steel Welds," *British Welding J.*, **8**, 526–540 (1961).
12. Borland, J. C., "Generalized Theory of Super-Solidus Cracking in Welds (and Castings)," *British Welding J.*, **7**, 508–512 (1960).
13. Borland, J. C. and Younger, R. N., "Some Aspects of Cracking in Welded Cr-Ni Austenitic Steels," *British Welding J.*, **7**, 22–59 (1960).
14. Wilkinson, F. J., Cottrell, C. L. M., and Huxley, H. V., "Calculating Hot Cracking Resistance of High Tensile Alloy Steel," *British Welding J.*, **5**, 557–562 (1958).
15. Reeve, L., *Trans. Inst. Welding*, **8**, 80–88 (1945).
16. Rollason, E. C. and Roberts, D. F. T., "An Explanation of Hot-Cracking of Mild Steel Welds," *Trans. Inst. Welding*, **13**, 129r–132r (1950).
17. Randall, M. D., Monroe, R. E., and Rippel, P. J., "Causes of Microcracking and Microporosity in Ultra-High Strength Steel Weld Metal," *Welding J.*, **41**, Res. Suppl., 193s–206s (1962).
18. Tremlett, H. F., "Unwanted Variation in Steel Composition. Effect on Welding," *J. West Scotland Iron Steel Inst.*, **70**, 90–100 (1962–1963).
19. Jones, P. W., "An Investigation of Hot Cracking in Low-Alloy Steel Welds," *British Welding J.*, **6**, 282–290 (1959).
20. Jones, P. W., "Hot Cracking of Mild Steel Welds," *British Welding J.*, **6**, 269–281 (1959).

21. Whiteley, J. H., "Apparent Relations Between Manganese and Segregation in Steel Ingots," *J. Iron Steel Inst.*, **2**, 63p–77p (1941).
22. British Welding Research Association, "Effect of Arsenic on Weldability and Notch Toughness of Mild Steel (To A. P. I. Grade B Specification), *British Welding J.*, **7**, 129–140 (1960).
23. Hirsch, W. and Fritze, H. W., "Über die Warmrissigkeit austenitischer Chrom-Nickel-Schweissen," *Schweissen und Schneiden*, **8**, 81–85 (1956).
24. Wiegand, H. H., "Schweissen hochwarmfester Stähle," *Schweissen und Schneiden*, **10**, 44–49 (1958).
25. Thomas, R. D., Jr., "Selection of Stainless Steel Electrodes for Trouble-Free Welds," *Metal Prog.*, **68**, 73–76 (1956).
26. Thielsch, H., "Alloying Elements in Chromium-Nickel Stainless Steels," *Welding J.*, **25**, Res. Suppl., 361s–404s (1950).
27. Thielsch, H., "Copper in Stainless Steels," Welding Research Council Bulletin No. 9 (Aug. 1951).
28. Medovar, B. J., "Good Weldable Austenitic Boron Steels for Ultra-High Parameter Steam Lines," *Welding in Power Plants,* International Institute of Welding Annual Assembly, Prague (1964).
29. Schaeffler, A. L., "Constitution Diagram for Stainless Steel Weld Metal," *Metal Progr.*, **56**, 680 and 680-b (1949).
30. Linnert, G. E., "Welding Type 347 Stainless Steel Piping and Tubing," Welding Research Council Bulletin No. 43 (Oct. 1958).
31. "Welding Handbook," Section 1, Fifth Ed., American Welding Society, New York, N.Y., 1962.
32. Hull, W. G. and Adams, D. F., "Fusion Welding of Aluminium Alloys. Part IV—Preliminary Tests on High-Strength Heat-Treatable Aluminum Alloys," *British Welding J.*, **1**, 513–521 (1954).
33. Mantle, E. C., "Welding with Aluminium Bronze," *Engineering*, **172**, 443 (1951).
34. Garriott, F. E., "Welding Iron-Bearing Alpha Aluminum Bronze," *Welding J.*, **31**, 18–28 (1952).
35. Lancaster, J. F., "The Cracking of Aluminium Bronze Welds," *British Welding J.*, **5**, 238–244 (1958).
36. Pease, G. R., "The Practical Welding Metallurgy of Nickel and High-Nickel Alloys," *Welding J.*, **36**, Res. Suppl., 330s–334s (1957).
37. Wilson, Jr., R. M., "Nickel Portion of A. W. S. 1950 Educational Lecture Series," *Welding J.*, **30**, 247–256 (1951).
38. Cordea, J. N., Kammer, P. A., and Martin, D. C., "Causes of Fissuring in Nickel-Base and Stainless Steel Alloy Weld Metals," *Welding J.*, **43**, Res. Suppl., 481s–491s (1964).
39. Interrante, C. G. and Stout, R. D., "Delayed Cracking in Steel Weldments," *Welding J.*, **43**, Res. Suppl., 145s–160s (1964).
40. Makara, A. M., Tsechal, V. A., and Zhovnitskii, I. P., "Determination of the Manner in Which Cold Cracks Develop in Welded Joints by Ultrasonic Defectoscopy," Avtomaticheskaya Svorka, **5**, 3–10 (1961).
41. Rollason, E. C. and Roberts, R. R., "Effect of Cooling Rate and Composition on the Embrittlement of Weld Metal," *J. Iron Steel Inst.*, **166**, 105–112 (1950).

42. Warren, W. G. and Vaughan, H. G., "The Initiation of Brittle Fracture at Welded Joints in Steel Structures," *Trans. Inst. of Welding,* **16**, 127–135 (1953).
43. Vaughan, H. G. and de Morton, M. E., "Hydrogen Embrittlement of Steel and Its Relation to Weld Metal Cracking," *British Welding J.,* **4**, 40–61 (1957).
44. Flanigan, A. E. and Miclev, T., "Relation of Preheating to Embrittlement and Microcracking in Mild Steel Welds," *Welding J.,* **32**, Res. Suppl., 99s–106s (1953).
45. Schaeffler, A. L., Campbell, H. C., and Thielsch, H., "Hydrogen in Mild-Steel Weld Metal," *Welding J.,* **31**, Res. Suppl., 283s–309s (1952).
46. Blake, P. D. and Pumphrey, W. I., "Effects of Hydrogen in Wrought Steel and Ferrous Weld Metal," *British Welding J.,* **6**, 211–224 (1959).
47. Jones, T. E. M., "Cracking of Low-Alloy Steel Weld Metal," *British Welding J.,* **6**, 315–323 (1959).
48. Masubuchi, K. and Martin, D. C., "Mechanisms of Cracking in HY-80 Steel Weldments," *Welding J.,* **41**, Res. Suppl., 375s–384s (1962).
49. Younger, R. N., Borland, J. C., and Baker, R. G., "Heat-Affected Zone Cracking of Two Austenitic Steels During Welding," *British Welding J.,* **8**, 575–578 (1961).
50. Younger, R. N. and Baker, R. G., "Heat-Affected Zone Cracking in Welded Austenitic Steels During Heat Treatment," *British Welding J.,* **8**, 579–587 (1961).
51. Williams, N. T. and Myers, J., "A Note on Sub-Surface Cracking of Austenitic Steels During Welding," *British Welding J.,* **9**, 432–435 (1962).
52. Stout, R. D. and Doty, W. D., "Weldability of Steels," Welding Research Council, New York, 1953.
53. Travis, R. E., Barry, J. M., Moffatt, W. G., and Adams, Jr., C. M., "Weld Cracking Under Hindered Contraction: Comparison of Welding Processes," *Welding J.,* **43**, Res. Suppl., 504s–513s (1964).
54. Carpenter, S. T. and Linsenmeyer, R. F., *Weld Flaw Evaluation,* Serial No. SSC-105 Final Report of Project SR-126 to the Ship Structure Committee, National Academy of Sciences, Washington, D.C. (July 29, 1958).
55. Warren, W. G., "Fatigue Tests on Defective Butt Welds," *Trans. Inst. Welding,* **15**, 112r–117r (1952).
56. Weck, R., "The Fatigue Problem in Welded Construction," Proc. Internat. Conf. on Fatigue of Metals, Inst. Mech. Engrs., London, pp. 704–717 (1956).
57. Ruttmann, W. and Baumann, K., "Erprobung und Bewährung austenitischer Schweissungen in der 610°C-Dampfkraftanlage," *Fachbuchreihe Schweisstechnik,* **3**, 16–26 (1955).
58. Ruttmann, W., Baumann, K., and Möhling, M., "Einige Erfahrungen mit austenitischen, insbesondere warmfesten Schweissungen," *Schweisstechnik* (Vienna), No. 1 and 2 (1957).
59. Ruttmann, W. and Brunzel, N., "10 Jahre austenitische Stähle im Kesselbetrieb," *Mitt. Ver. Grosskesselbesitzer,* **80**, 310–326 (Oct. 1962).
60. Gunn, K. W. and McLester, R., "Effect of Mean Stress on Fatigue Properties of Aluminium Alloy Butt-Welded Joints," *British Welding J.,* **7**, 201–207 (1960).

61. Newman, R. P., "The Influence of Weld Faults on Fatigue Strength with Reference to Butt Joints in Pipe Lines," *Trans. Inst. Marine Engrs.*, **68**, 153–172 (1956).
62. Dinsdale, W. O. and Young, J. G., "Significance of Defects in Aluminium Alloy Fusion Welds," *British Welding J.*, **9**, 482–493 (1962).
63. Wilson, W. M., Munse, W. H., and Snyder, I. S., "Fatigue Strength of Various Types of Butt Welds Connecting Steel Plates," Univ. of Illinois Eng. Exper. Station, Bulletin No. 384 (Mar. 1950).
64. Hartman, A. J., Bruckner, W. H., Mooney, J., and Munse, W. H., "Effect of Weld Flaws on the Fatigue Behavior of Butt-Welded Joints in HY-80 Steel," University of Illinois, Urbana, Illinois (Dec. 1963).
65. British Welding Research Association, "Faults in Arc Welds in Mild and Low Alloy Steels," *Trans. Inst. Welding*, **13**, 3r–15r (1950).
66. Warren, D. and Stout, R. D., "A Review of the Literature on Porosity in Mild Steel Weld Metal," *Welding J.*, **31**, Res. Suppl., 381s–386s (1952).
67. Muir, J., "Review of Literature on Porosity," *British Welding J.*, **3**, 98–102 (1956).
68. Bradley, J. W. and McCauley, R. B., "The Effects of Porosity in Quenched and Tempered Steel," *Welding J.*, **43**, Res. Suppl., 408s–414s (1964).
69. Adams, D. F., "Strength of Metal-Arc Welds in HP 30 Aluminium Alloy, *British Welding J.*, **5**, 568–575 (1958).
70. Newman, R. P., "Effect on Fatigue Strength of Internal Defects in Welded Joints," *British Welding J.*, **6**, 659–664 (1959).
71. Homes, G. A., *Arcos*, **15**, 1951–1957 (1948).
72. Masi, O. and Erra, A., "Radiographic Examination of Welds. A Complete Assessment of Defects in Terms of Tensile and Fatigue Strength," *Metallurgia Ital.*, **45**, 273–283 (1953).
73. Hempel, M. and Möeller, *Archiv. Eisenhütten.*, **20**, 375–383 (1949).
74. Kihara, H., Tada, Y., Watanabe, M., and Ishii, Y., "Nondestructive Testing of Welds and Their Strength," Society of Naval Architects of Japan, Tokyo, Japan.
75. Hempel, M., *Stahl u. Eisen*, **58**, 756–760 (1938).
76. Kauczor, E., "Brüche durch unsachgemässes Schweissen," *Schweissen und Schneiden*, **15**, 430–434 (1963).
77. Dinsdale, W. O. and Young, J. G., "Tungsten Inclusions in $\frac{1}{4}$ in. Thick Al-Mg-Mn Alloy NP8," *British Welding J.*, **11**, 238–244 (1964).
78. Private Communication, Bayernwerk, A. G.
79. Bureau of Ships, U.S. Navy, "Fabrication, Welding and Inspection of HY-80 Submarine Hulls," NAVSHIPS 250-637-3 (Jan. 1962).
80. Wilson, W. M., Bruckner, W. H., McCrackin, Jr., T. H., and Beede, H. C., "Fatigue Tests of Commercial Butt Welds in Structural Steel Plates," Univ. of Illinois Eng. Exper. Station Bulletin No. 344 (Oct. 12, 1943).
81. Wood, J. L., "Flexural Fatigue Strength of Butt Welds in NP5/6 Type Aluminium Alloys," *British Welding J.*, **7**, 365–380 (1960).
82. Dinsdale, W. O., "Effect of Reinforcement Shape on Fatigue Behavior of Butt Welds in NP5/6," *British Welding J.*, **11**, 233–238 (1964).
83. Lea, F. C. and Parker, C. F., "Reports on Physical Tests," *Proc. Inst. Mech. Engrs.*, **133**, 15–63 (1936).

84. Lea, F. C. and Whitman, J. G., "The Failure of Girders Under Repeated Stresses," *J. Inst. Civil Engrs.*, **9**, 301–328 (1937–38).
85. Fall, H. W., Brugioni, D. L., Randall, M. D., and Monroe, R. E., "Improvement of Low-Cycle Fatigue Strength of High-Strength Steel Weldments," *Welding J.*, **41**, Res. Suppl., 145s–153s (1962).

———chapter 13

EFFECTS OF WELDING ON DEFECTS AND FAILURES

COMMONLY USED WELDING PROCESSES

Forty years ago, pressure vessels and pipe were welded by the oxyacetylene process employing bare wire as filler metal. Of course, applications generally were not critical and little, if any, inspection was required.

In the nineteen-thirties, the shielded metal-arc process was applied to pressure vessel and pipe welding. Good quality welds were readily produced where welding was possible from both sides. In many pipe butt welds this was not practical as the pipe inside generally is not accessible for welding in diameters of less than 24 in., and is often inaccessible even in larger diameters. As quality requirements in pipe welds became more critical as, for example, in steam power plant applications, backing rings were employed. Initially, all backing rings were flat. As vessel and pipe diameters and wall thicknesses increased, however, problems arose in fit-up that tended to enhance cracking. To overcome this, the inside pipe ends were taper machined and taper machined backing rings were employed to facilitate fit-up and improve the quality of the weld in the root of the pipe joint.

In the late forties, the inert-gas tungsten-arc welding process was applied to pipe welding. The primary advantage of this method was to the elimination of backing rings. On light wall pipe, the entire butt weld is then made by inert-gas tungsten-arc welding. On heavier wall pipe, only the first (or root) pass, or the first two passes, are made in this manner. The balance of the weld is generally completed subsequently by shielded metal-arc welding with coated electrodes or by submerged-arc welding.

The submerged-arc process is extensively used for welding seam and circumferential butt joints in pressure vessels. This process is also used in shop pipe welding where the pipe can be rotated.

In recent years, the gas-shielded consumable metal-arc processes have been used increasingly, particularly in pipe welding.

It is extremely important to recognize that each welding process tends to produce different types of defect conditions. Some of these may be quite hazardous and yet rather difficult to detect by the major common nondestructive testing methods.

Actually, there is no welding process that guarantees soundness in pressure vessel or pipe welds. For each application, techniques should be carefully evaluated under conditions identical with, or even more critical than, those that will apply to the production weld. A weld that is sound when made on a relatively short test section may contain cracks or other defects when made under conditions of restraint.

WELDING PROCESS CONSIDERATIONS

Submerged-Arc Welding

High quality welds in pressure vessels and piping can be produced by the submerged-arc welding process. Nevertheless, lack of experience with the method has resulted in defective welds. Filler metal and flux compositions may also affect weld soundness. Cracking in weld passes can occur if the freezing conditions of the weld bead are such that weld shrinkage causes cracking. This has occurred in circumferential butt welds, Fig. 13-1, as well as in longitudinal seam welds, Fig. 13-2. These angular cracks are sometimes very difficult to detect by nondestructive testing techniques.

Inert-Gas Tungsten-Arc Welding

While the advantages of the inert-gas tungsten-arc welding technique were quickly promoted, the accompanying disadvantages were

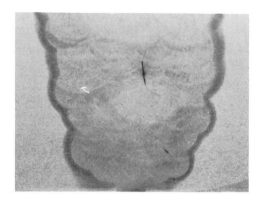

Fig. 13-1. Cracking in two weld beads in circumferential pipe butt joint made by submerged-arc welding process in $1\frac{1}{4}$ Cr–$\frac{1}{2}$ Mo alloy steel piping materials. (Magnification 3×)

not so readily recognized. Shielded metal-arc welding of the first weld pass in pipe butt joints against backing rings results in a relatively heavy weld deposit that is supported by the backing ring. Inert-gas tungsten-arc root-pass welding results in a much thinner weld deposit. Whereas the weld deposit made by shielded metal-arc welding freezes toward the pipe ends and the backing ring, the inert-gas tungsten-arc root-pass weld freezes only towards the pipe ends. This condition, along with the thinner cross section, increases significantly the tendency toward cracking in the center of the root weld pass.

Unfortunately, cracks in the root of a weld are frequently not detected if radiographic inspection is made only of the completed weld. The difficulty of detecting cracks in the initial weld root pass, even by X-Ray radiography, is illustrated in Fig. 13-3. The root in the 8-in. Schedule 160 (0.906-in. wall) $1\frac{1}{4}$Cr–$\frac{1}{2}$Mo alloy steel pipe which was made by the inert-gas tungsten-arc welding process, contained the $2\frac{1}{2}$-in. long crack shown in Fig. 13-3. As is standard practice, the subsequent passes were made by shielded metal-arc welding with $1\frac{1}{4}$Cr–$\frac{1}{2}$Mo (AWS-ASTM E8016-B2) coated "stick" electrodes. After the third cover pass with the coated electrode, the root pass crack was no longer visible by X-Ray radiography, as shown in Fig.

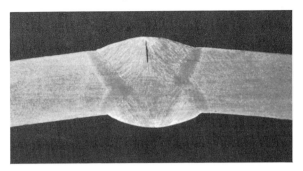

Fig. 13-2. Cracking in longitudinal seam weld made by submerged-arc tandem-arc method on carbon steel pipe. (Magnification 1×)

13-3(D). Subsequent sectioning of the completed weld, which consisted of nearly 10 shielded metal-arc welding passes, revealed nevertheless that the root pass crack was still present, Fig. 13-3(H).

The "disappearance" of the crack was caused by shrinkage of the first coated electrode pass made on the inert-gas tungsten-arc root pass, which drew the crack together tightly. Certainly, the potential danger of such cracks has not been lessened by their being closed up mechanically.

Detection of these cracks in chromium-molybdenum alloy steel welds is further complicated by the continuous preheat requirements of most specifications. Typically, these state that the preheat temperature (usually between 500 and 600°F) must be maintained without interruption until the whole weld has been completed, or until at least several layers of weld metal have been deposited.

For example, PFI Standard ES-8 [1] states that "welding on 1Cr–½Mo to 3Cr–1Mo materials may be interrupted at any time provided that a minimum of at least ⅜ in. thickness of weld deposit or 25% of the welding groove is filled, whichever is greater." On 5Cr–½Mo or higher alloy steels, welding should not be interrupted unless a partial or complete stress-relieving heat treatment is given.

Inspection of the root pass weld or on heavier wall piping, of the second or third weld pass while at preheat temperature, by radiographic techniques would require "hot radiographic" water-cooled film holding fixtures.

276   *Defects and Failures in Pressure Vessels and Piping*

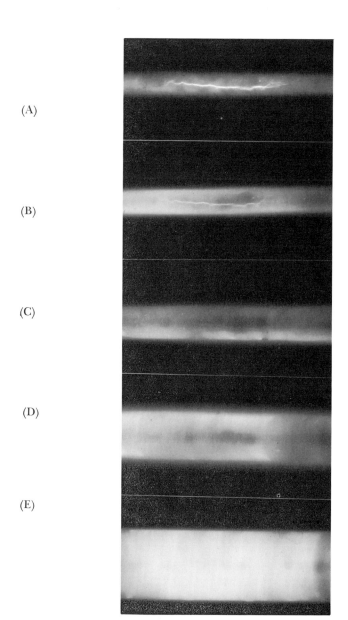

(A)

(B)

(C)

(D)

(E)

Fig. 13.3. See opposite page.

Effects of Welding on Defects and Failures 277

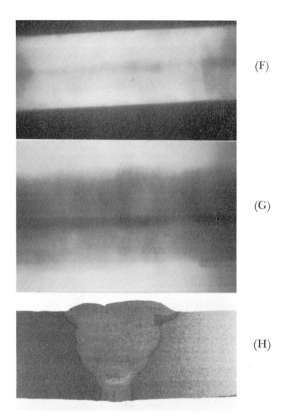

Fig. 13.3. Crack in root pass of 1¼ Cr–½ Mo pipe butt weld made by inert-gas tungsten-arc welding. (A) Radiograph of root pass containing 2½-in. long centerline crack. (D) through (G) crack no longer visible on radiographs after third and subsequent weld layers made by shielded metal-arc welding with covered electrodes. (H) Cross section illustrating presence of crack in completed weld. (Magnification 1×)

Gas-Shielded Consumable Metal-Arc Welding

Consumable metal-arc welding under the protection of CO_2, argon, or helium is increasingly employed in piping applications. Although this process produces welds adequate for many noncritical services, it does not produce consistently high quality butt welds. Such welds are susceptible to weld craters, cracks, or lack of fusion in the root pass.

278   *Defects and Failures in Pressure Vessels and Piping*

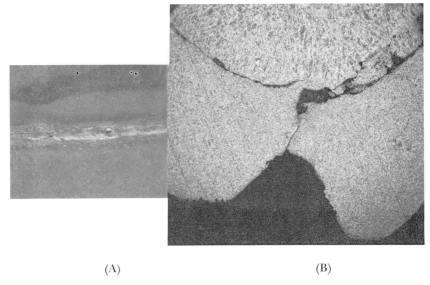

(A)   (B)

Fig. 13-4. Weld crater visible on underside of carbon steel pipe weld made by consumable metal-arc welding. Sectioning of weld across crater revealed incomplete fusion between weld beads at root of weld joint. (Magnification of (B) 25×)

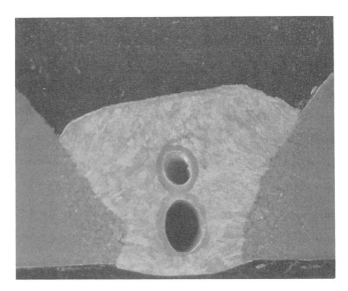

Fig. 13-5. Cross section of root pass in carbon steel pipe weld which contained "worm holes." Weld was made by consumable metal-arc welding. (Magnification 8×)

Effects of Welding on Defects and Failures

Fig. 13-6. Liquid penetrant indication of incomplete fusion along pipe inside.

With the equipment and techniques presently employed, it is difficult to avoid craters in the weld root adjacent to the location of the tack welds as shown in Fig. 13-4. Worm holes may also result as shown in Fig. 13-5.

Incomplete fusion in the root pass and between weld passes is often not readily detectable on radiographic films. Liquid penetrant examination of the weld root, shown in Fig. 13-6; bend tests, illustrated in Fig. 13-7; or cross sections, shown in Fig. 13-8 reveal this type of defect more effectively.

Flash and Resistance Welding

Flash and resistance butt welding of boiler tubing can also result in problems. For example, insufficient gas protection against oxidation and scaling can cause improper fusion of the tube ends.

Weld flash remaining on the inside of joints can permit entrapment of corrosives or introduce other conditions enhancing localized

280 *Defects and Failures in Pressure Vessels and Piping*

Fig. 13-7. Examples of cracking in root bend tests of welds made by gas-shielded metal-arc welding process.

Fig. 13.8. Etched cross section of incomplete fusion in root of weld made by gas-shielded metal-arc welding process. (Magnification 5×)

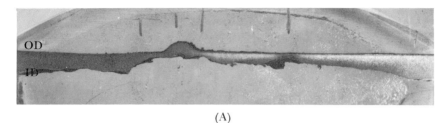

Fig. 13.9. Corrosion failures in boiler tube butt welds where weld metal remained on tube inside (B). On fireside of tube, corrosion occurred in area between metal flash and tube wall (A).

corrosion, particularly on the fireside of the boiler tubing. This is illustrated in Fig. 13-9.

Improper flash welding of 2¼-in. OD by 0.2-in. wall riser tubing of mild steel resulted in welds containing slag and oxide layers causing leaks revealed in pressure testing of a new boiler. Bend tests across the welds failed at angles of less than 30 degrees.[2]

WELDING PROCEDURE CONSIDERATIONS

Welding Filler Metal Controls

Some inexperienced fabricators still use either "all-purpose" electrodes or purchase surplus electrodes with little regard to composition. Many of the so-called all-purpose electrodes sold to "weld all steels" actually are Type 312 (29Cr–9Ni) or Type 310 (25Cr–20Ni) stainless steel electrodes. Though they can be readily deposited on alloy steel materials, the properties of the weld deposit may differ sufficiently from a lower alloy base metal to cause failure, particularly if the service involves mechanical or thermal fatigue at high temperatures. For example, in a steam power plant, cracking occurred in a

282 *Defects and Failures in Pressure Vessels and Piping*

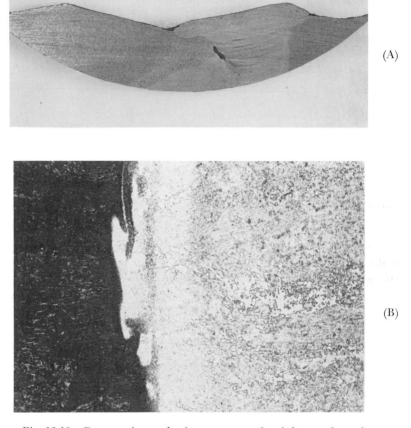

Fig. 13-10. Cross section and microstructure of stainless steel repair weld (left in B) in ½ Mo steel pipe material. [Magnifications of (A) 1×, (B) 100×]

valve end after service for nearly 10 years at 900°F. Originally, it had been repair welded with a stainless steel electrode. The stainless steel weld area absorbed carbon from the adjacent ½ Mo steel butt weld deposit which became brittle and subsequently failed. A cross section of the weld joint outside the failed area is shown in Fig. 13-10. The "black" area represents the stainless steel weld deposit at its furthest extension.

Effects of Welding on Defects and Failures 283

(A) (B)

Fig. 13-11. Cracking in ½ Mo steel weld deposit made 1 Cr–½ Mo steel base metal. (A) reduced to ½ size (B) Magnification 6×.

As pointed out elsewhere, care must be taken to insure that the correct welding filler metals are used. Mixups in welding filler metals in the welding shop can lead to service failures in many critical applications.

Figures 13-11 and 13-12 illustrate a developing failure in a joint in which 1Cr–½Mo base metal was welded with ½Mo steel electrodes, even though 1¼Cr–½Mo electrodes had been specified. The joint had been in service for approximately two years at 1000°F. The carbon migrated from the weld deposit into the base metal, leaving a decarburized zone in the weld deposit adjacent to the bond as shown in Fig. 13-12A.

Similar instances involving cracking and failures of weld deposits made with welding filler metals other than those specified have also been reported elsewhere.

A steam header 83 in. in diameter by 2.32-in. wall made of a low alloy steel containing 0.17% C and Cu, Ni and Mo required the welding of various nipples of mild steel. Welding electrodes of CuNi steel had been specified. Instead, the fabricator used 2½Cr steel electrodes. After welding, the header was stress relieved. Subsquent pressure testing revealed leaks in seven nipples, caused by star-type cracks in the welds. Cracking in the leaking and other joints occurred in the weld deposits only. It was associated with excessive hardening produced by depositing the 2½Cr electrode on the CuNiMo steel at a preheat of 210 to 300°F. This temperature was considered too low for the 2½Cr weld that was further hardened by Ni and Cu from the base metal. Because of dilution, the actual Cr content of the weld was approximately 1%.[3]

A number of failures have been reported also in refinery applica-

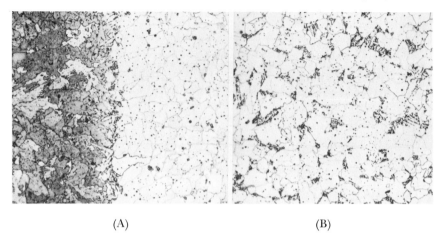

(A) (B)

Fig. 13-12. Photomicrographs of (A) "bond" zone between decarburized ½ Mo steel weld metal and 1 Cr–½ Mo steel base metal. (B) Microstructure of ½ Mo steel weld metal away from the decarburized zone. (Magnifications 300×)

tions involving catalytic reformers of chromium-molybdenum alloy steels which had been welded with carbon steel electrodes. The resulting failures have been caused by hydrogen embrittlement. In one catalytic reformer operating at 950°F and 375 psi, it was found that the longitudinal seam welds in six 1¼Cr–½Mo heat exchangers had been made with carbon steel. The weld deposit contained only 0.24Cr and 0.13Mo and had become thoroughly decarburized. Two of the welds had cracked and one leaked.[4]

In another instance, one of three companion catalytic reformers of 1Cr–½Mo alloy steel developed one or more ruptured methane blisters on the inside surface of seam welds. Metallographic examination revealed intergrannular pitting, decarburization and fissuring typical of hydrogen embrittlement. The alloy content of the welds varied between 0.02 and 0.54Cr. The reactors had operated at 880 to 900°F and 277 to 405 psi hydrogen partial pressure for four years.[5]

Inert-Gas Tungsten-Arc Welding

In the inert-gas tungsten-arc welding of light wall stainless steel pipe or tubing, a rough, heavy oxide scale is frequently produced on the underside of the joint. This is illustrated in Fig. 13-13. To avoid this scale formation, the best procedure is to introduce a so-called

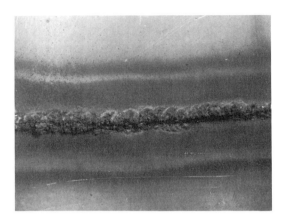

Fig. 13-13. Heavy "cauliflower-type" oxide scale on underside of stainless steel pipe weld.

purging gas such as argon, helium, or nitrogen on the underside of the joint. By effectively displacing the air, a sound and clean weld joint is obtained as shown in Fig. 13-14.

For large diameter piping, another procedure is to grind away the heavy scale and then "back weld" from the inside. Although the resulting weld quality may not be as good as that of a pipe weld made by purging, it will be acceptable in the majority of applications.

In the grinding and back welding technique, it is frequently difficult to ascertain that all of the scale has been ground away and that the new groove made by grinding to prepare the joint for back welding is centered above the joint. If the groove is too narrow, there is a tendency to offset the groove from the weld center and leave an undesirable lack-of-penetration area. Since the grinding marks tend to smear over the lack-of-penetration area, this offset might not be readily noted.

Because grinding is costly, some welding shops eliminate the grinding operation. Back welding is subsequently done by inert-gas tungsten-arc welding over the heavily scaled area. This, however, is poor practice since it tends to bury the scale and may result in cracking as illustrated in Fig. 13-15. Since these cracks may not penetrate to the outer surface, they are generally not detected. In fact, the visual appearance of such welds may be as clean as one that is prop-

286 *Defects and Failures in Pressure Vessels and Piping*

Fig. 13-14. Smooth clean weld surface on underside of stainless steel pipe weld where the air has been effectively displaced (purged) with argon.

erly ground and back welded or even one that is purged. During subsequent corrosive service, however, these scale (oxide) filled cracks may open up and eventually permit leakage of the corrosive solution. Mechanical or thermal fatigue also can open these types of cracks.

Occasionally, special proprietary backup fluxes are painted on the underside of a joint to protect the metal from oxidation. Unfortunately, almost all of these fluxes become extremely adherent to the underside of the weld. Since even wire brushing is likely to be insufficient to remove flux scale, grinding is usually necessary. In several instances the flux scale has also been "removed" by back welding. This may produce a clean appearing but defective weld because the flux becomes entrapped in the weld, as shown in Fig. 13-16 which represents a commercial mitered elbow.

Since visual inspection is frequently the only criterion for acceptance of light wall stainless steel piping and vessels, a purchaser viewing a smooth weld surface may hold the comforting but erroneous belief that his welds are sound and of a quality comparable to that of the base metal.

Submerged-Arc Welding

On alloy steel piping, good welding practice should involve the use of alloyed steel welding filler wires of a composition correspond-

Effects of Welding on Defects and Failures

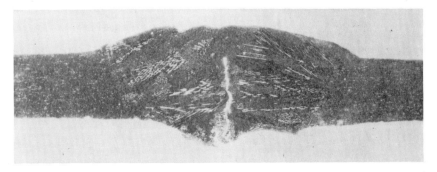

Fig. 13-15. Incomplete penetration (or cracks) caused by scale embedded inside in stainless steel pipe weld joint where back welding was done from the inside of the pipe. (Magnification 12×)

ing to that of the base metal with a neutral flux. The use of fluxes containing metallic alloying elements in the form of powders with carbon steel filler wires generally is not considered good practice. If alloyed fluxes are used, the alloy content of the weld deposit tends to increase with increasing weld interpass temperatures. For example, on $2\frac{1}{4}$Cr–1Mo pipe, the weld deposit made with corresponding alloyed fluxes may contain more than 5% chromium when the interpass temperature increases to 700 to 800°F.

Columbium-Bearing Stainless Steels

The welding of columbium-bearing stainless steels frequently presents problems in the weld deposit and heat-affected zone, due to their tendency to hot cracking.[6] This is particularly true in heavy sections.

In *weld deposits*, the fissuring is associated closely with composition, particularly as it affects the formation of (delta) ferrite. In ferrite-free fully austenitic weld microstructures, fissuring can usually be expected. When small amounts of ferrite are formed, fissuring is reduced. Normally, about 3 to 8% ferrite is desired in the microstructure. In Type 347 compositions, welding filler metals are normally made of compositions resulting in weld deposits containing 3 to 8% ferrite.

Major cracking can also occur in columbium-bearing stainless steels, particularly when welding under restraint. Such cracks, of course, require repair. As discussed in Chapter 12, it is important

Fig. 13-16. Flux trapped in weld in stainless steel mitered elbow. Weld was made by inert-gas tungsten-arc welding from one side. The back-up flux used for protection from the air of the underside of the stainless steel weld was trapped in the weld. (Magnification 12×)

to differentiate between major cracking and fissuring in weld deposits.

Whereas most types of stainless steels can be readily joined without the addition of filler metal, in the columbium-bearing grades it is almost always necessary to utilize welding filler metals to avoid major cracking and minimize microfissuring.

In the *base metal* of columbium-bearing stainless steels, cracking may occur immediately after welding or after subsequent high temperature stress-relief heat treatments. For example, a vessel being fabricated for a nuclear reactor from Type 347 stainless steel plates exhibited microfissuring in the heat-affected zones of a few plates. The cracks which became evident after postweld heat treatment resulted from high residual stresses caused by restraint conditions imposed by design and fabrication.[7]

Another nuclear reactor vessel of Type 347 stainless steel materials involved the welding of nipples into the shell. During heat treatment at 1850°F, stress-rupture cracking of the shell plate material occurred adjacent to the welds. The repair procedure reducing residual stresses involved the removal of the nipples, including the welds and cracked heat-affected zones. Only alternate nipples were welded into place,

Effects of Welding on Defects and Failures

after which the vessel was heat treated. Subsequently, the remaining nipples were welded into place and the vessel was heat treated again. With this procedure, apparent cracking difficulties were avoided.[7]

Where these steels are used in service other than corrosive service, base-metal cracking is also sometimes minimized by welding with special welding filler metals which exhibit increased ductility in weld deposits at elevated temperatures. One of the filler metals used is a stainless steel containing a 16Cr–8Ni–2Mo type composition. The higher ductility at elevated temperature tends to reduce the straining of the base metal during cooling, thus minimizing cracking. These "high-temperature" ductility welding filler metals also help to minimize fissuring during postheat treatments, where these are required. Moreover, when heat treatments are required, full furnace heat treatment is far more beneficial than localized heating.

Dissimilar Metal Joints

In addition to the importance of proper filler metal selection discussed in Chapter 4, welding of dissimilar metal joints must be done under carefully controlled conditions. This includes preheat and postheat treatments.

A failure occurred for example in a $2\frac{1}{4}$Cr–1Mo pipe welded with 18Cr–8Ni–2Mo (Type 316) stainless steel electrodes. The welding was done without preheat. Because of the excessive hardness which resulted in the base metal, the weld was subsequently stress relieved by induction heating at 1435°F. During cooling, severe cracking was evident along the base metal to weld metal fusion zone.[8] The metallurgical structure of such welds reveals severe decarburization in the heat-affected zone of the ferritic alloy steel.[9,10]

Titanium Vessels and Pipe

Cleaning and purging are essential when welding titanium, zirconium, and other materials that are highly susceptible to the absorption of oxygen, nitrogen, and other gaseous contaminants. This contamination may result in weld and heat-affected zone areas that are inherently extremely brittle, although the surface appears sound. Since severe cleanliness techniques may double the actual welding costs, many welds are not properly made.

A typical failure by cracking of a weld in a titanium coil is shown

290   *Defects and Failures in Pressure Vessels and Piping*

Fig. 13-17. Cracking of butt weld in titanium tubing after several months of chemical plant service. Weld was brittle as a result of improper purging.

in Fig. 13-17. Some argon purging had been done on the inside, but although most of the air was displaced, enough oxygen and nitrogen remained to embrittle the weld joint.

Titanium sheet is also used to line steel vessels and pipe and steel or cast iron valves exposed to highly corrosive service conditions. Improperly made weld joints which are either brittle or contain fine cracks may fail in service. Figure 13-18 illustrates a failure in a 10-in. titanium lined cast iron valve body. After the initially brittle weld had failed in the sodium chlorate service at 140°F, the cast iron body was corroded out within a few days.

FAILURE CONDITIONS

Alignment Welds

To facilitate alignment, particularly on piping, clips or channels are at times welded across the ends of two pipe sections. This repre-

Fig. 13-18. Failure in a titanium lined valve body in sodium chlorate service caused by brittle weld in titanium lining.

292 *Defects and Failures in Pressure Vessels and Piping*

sents a common practice in field erection of carbon and low-alloy steel piping.

After the weld between the two pipe sections has been started, the alignment clips are usually knocked off or removed.

The effect of these surface welds generally is similar to arc strikes.

Where the adjacent welds between the two pipe ends are subsequently stress relieved, the harmful effect of these surface welds is generally removed.

In a few instances, subsequent cracking has been observed.[3]

Lack of Penetration

Lack of penetration has been involved in a significant number of failures of pressure vessel, tank and pipe welds.

Figure 13-19 illustrates rupture of a steam boiler shell. The failure, which was nearly fatal, was primarily due to a severe lack of penetration in the circumferential butt welds. A cross section of the weld

Fig. 13-19. Service failure of circumferential butt weld in boiler shell certified to ASME Boiler and Pressure Vessel Code.

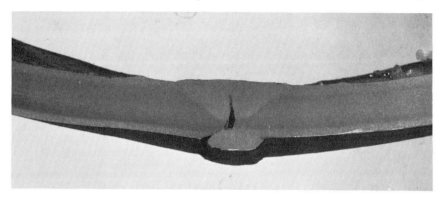

Fig. 13-20. Cross section of circumferential butt weld in area of boiler that did not fail.

at a location where failure had not yet occurred is illustrated in Fig. 13-20.

Sometimes, the failure process caused by lack of penetration in a weld may be somewhat complex. Figure 13-21 illustrates a tube failure in the vacuum section of a heat exchanger. The $3/8$-in. thick plate separating the vacuum side from the steam side was welded from one side only. During service involving some thermal fatigue, the lack of penetration in the weld resulted in cracking across the plate weld. This allowed passage of steam through the crack to the vacuum side. The steam impinging on the $3/4$-in. OD copper tubes caused erosion of the tubes (see hole at "A"). Once the hole was produced in the tube, a water spray hitting the steel plate caused erosion of a hole through the cracked weld in the steel plate (indicated at "B").

In December 1951, a 1115-ft diameter by 30-ft high water storage tank failed in New Mexico. Subsequent investigation revealed a faulty butt weld in the $1/2$-in. thick plate. The plate had been flame cut and had received no edge preparation before rewelding. In some areas, the weld penetration was less than 0.1 in. The tear propagated through the solid plate. The steel itself was of good quality.[11]

Lack of penetration in submerged-arc welds has been revealed during hydrostatic testing. For example, in a seam welded pipe section containing a double submerged-arc weld, failure occurred where the inside weld bead had missed the plate edge for about 8 in.[12]

294 *Defects and Failures in Pressure Vessels and Piping*

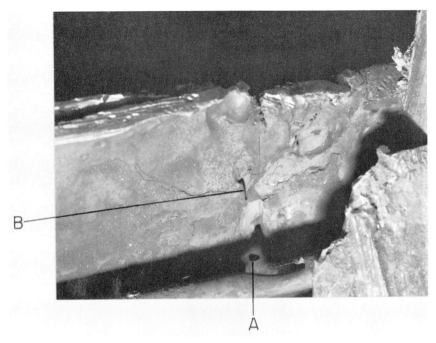

Fig. 13-21. Failure of heat-exchanger tubing caused by lack of penetration in $3/8$-in. thick plate butt weld separating vacuum side from steam side.

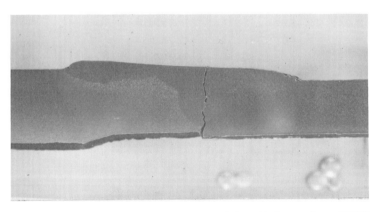

Fig. 13-22. Failure in steel tank containing 40% caustic at 240°F caused by lack of penetration in weld.

Effects of Welding on Defects and Failures

Lack of root penetration and poor fit-up were the principal causes of an 8-in. long rupture in the circumferential butt weld of a pressure reducing section operating at 930/570°F and 1800/320 psi.[13] Similar failures and cracking have been reported in a number of other steam piping systems.[14]

In some corrosive environments, good fit-up and complete weld penetration may be particularly important. Figure 13-22 illustrates failure in a weld joint of a steel tank containing 40% caustic at 240°F after one month of service. Welds showing proper penetration did not fail.

Arc Strikes

When the service involves thermal fatigue, arc strikes may lead to cracking in the heat-affected zones adjacent to the weld area as shown

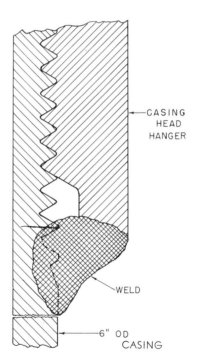

Fig. 13-23. Location of cracks in 6-in. OD casing caused by seal welding of high-carbon pipe casing steel.[15]

296 *Defects and Failures in Pressure Vessels and Piping*

in Fig. 12-4. When cracking occurs it generally can be related to the higher hardness produced in these zones. Generally these cracks do not penetrate very deeply into the vessel or pipe base metal, and often are not propagated beyond a certain limit. In many applications the presence of light arc strikes may not be considered critical. They can be rendered even less harmful by post weld stress-relief heat treatment which softens the heat-affected zone. Grinding to remove the notch effect also may be beneficial but may not prevent cracking.

Hardness Variations

As discussed in other Chapters, hardness variations of significant magnitude may cause failure.

Deposition of a heavy seal weld bead on a 6-in. OD high carbon steel casing around the bottom of a hanger to provide a positive seal against leakage resulted in severe cracking, Fig. 13-23. Failure of the N-80 casing material occurred about an hour after welding while cir-

Fig. 13-24. Porosity in weld deposit made by gas-shielded consumable metal-arc welding caused by inclusions in steel base metal.

Effects of Welding on Defects and Failures

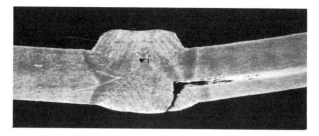

Fig. 13-25. Cracking in weld deposit made by submerged-arc welding caused by lamination in steel base metal.

culating for cementing. Welding increased the normal hardness of the casing from about 255 to 515 Brinell in the heat-affected zone.[15]

Base-Metal Inclusions

Base-metal inclusions will affect differently the weld quality produced by different welding processes. They affect most adversely weld deposits made by the gas-shielded arc-welding processes. Figure 13-24 illustrates porosity caused in welds adjacent to "dirty" steel welded by gas-shielded consumable metal-arc welding. Shielded metal-arc welding of the same steels fluxed out the inclusions and minimized weld defects to acceptable levels.

The beneficial effect of the flux on coated electrodes and in submerged-arc welding decreases with increases of the weld bead size and welding current.

Cracking caused by a severe lamination in steel plate welded by the submerged-arc welding process is illustrated in Fig. 13-25. The weld was made in two passes utilizing a tandem arc at current levels of about 800 amps.

WELDING OF EQUIPMENT IN SERVICE

Sometimes it is desirable or necessary to weld equipment which has already been in service. Wherever possible, such welding should be done during shutdown periods to insure proper weld end preparation, cleaning, and inspection.

Occasionally, this is not possible. Then welding may have to be done on the vessel or piping component containing liquids or gases. This can be dangerous and has led to a number of serious explosions. However, even when explosions are not likely, improper planning and lack of recognition of metallurgical considerations have resulted in severe damage and led to subsequent service failures.

For example, handling lugs were oxyacetylene welded onto propane bottles. Although the valves on the bottles were opened, some gas apparently remained within the bottles. Welding to the surface of the 0.08-in. thick bottles resulted in severe carburization of the inside surface. This lowered the melting point of the carbon steel and even caused localized surface melting. After cooling, an extremely hard white-cast-iron type surface layer developed in the heated area resulting in cracking.[16]

REPAIR OF WELD DEFECTS

Many defects in weld deposits in pressure vessels, tanks, and piping can be removed by grinding, chipping, machining, or flame or arc gouging. Properly made weld deposits to fill up the resulting cavity may return the material to normal soundness.

The repair by welding of defective or cracked parts should always be approached with caution. It happens, not infrequently, that a structure contains more serious defects after repair than it did before. Thus carefully planned procedures and proper welding techniques are necessary. Heat treatments, when required, also should be applied more carefully and uniformly.

There are many instances where a noncritical defect (which would have been extremely unlikely to cause failure) has been removed and the new weld contained cracks, craters or other defects which in time propagated across the wall thickness.

Vessels and piping that have been in service may contain fatigue cracks which represent severe stress raisers. In steels, the main danger associated with these defects is their tendency to start brittle failure. However, in many applications a fatigue crack may not be dangerous as long as the crack is under control. Cracks and other discontinuities are present in most structures. They are not dangerous if their length is insufficient to start static fracture, and if the service conditions do not involve brittle behavior.[17] Under such conditions, repair can do more harm than good.

Repair of defects is particularly critical when it may leave new defects at or near the inside surface of the vessel, tank, or pipe.

During hydrostatic testing in 1950 at 200 psi, cracking occurred in a 15-ft diameter pipe utilized as a penstock. The crack extended for 50 ft and traversed three pipe sections having plate thicknesses of $1\tfrac{5}{16}$ and $1\tfrac{9}{16}$ in. Examination indicated that the fracture started at a repair weld in a tunnel-welded girth joint. Heavy irregular weld beads applied during repair appeared to have caused a notch condition of significant magnitude.[18]

REPAIR OF FAILURES

Where pressure vessels, tanks or piping have been in service and have developed cracks or ruptures, extreme care is necessary to insure that any subsequent repairs are carefully made. This should include a thorough determination of the extent of cracking and of any permanent damage to the materials.

A 31-in. diameter by 66-in. long cylindrical compressed-air tank had developed cracking in the weld between the flat bottom plate and the shell. The subsequent weld repair did not involve complete removal of the defective weld. Although the vessel after welding passed a pressure test at 176 psi, it failed after being placed in service. The heavy 80-lb bottom plate ruptured suddenly along the entire weld and flew off, killing a workman standing some 23 ft away.[19]

REFERENCES

1. Standard ES-8, "Preheat and Postheat Welding Practices for Low Chromium-Molybdenum Steel Pipe," Pipe Fabrication Institute, Pittsburgh (Mar. 1964).
2. Verein. d. Grosskesselbesitzer, 1960/61 Review.
3. Verein. d. Grosskesselbesitzer, 1961/62 Review.
4. Manuel, R. W., "Hydrogen Service Failures of Welds With Insufficient Alloy Content," *Corrosion,* **17**, 435t–436t (1961).
5. Cooper, C. M., "Material Deficiencies in Welds of Hot Hydrogen Process Equipment Causes Serious Failures," *Materials Protection,* **3**, 34–40 (Jan. 1964).
6. Linnert, G., "Welding of 347 Stainless Steel Piping and Tubing," Welding Research Council Bulletin No. 43 (1958).

7. Slaughter, G. M., Private Communication, Oak Ridge National Laboratory, Oak Ridge, Tenn.
8. Baerlecken, E., "Schweissverbindungen zwischen austenitischen und ferritischen Stählen," *Mitt. Ver. Grosskesselbesitzer,* **26,** 529–540 (1953).
9. Kauczor, E., "Karbidzonenbildung bei austenitischen Schweissen an niedriglegierten und unlegierten Stählen," *Schweissen und Schneiden,* **14,** 482–485 (1962).
10. Erdmann-Jesnitzer, F., Beckert, M., and Schmiedel, H., "Gefügebedingte Schäden temperaturbeanspruchter austenitischer Schweissnähte für ferritische Stähle," *Schweissen und Schneiden,* **9,** 407–414 (1957).
11. "The Tucumcari Tank Failure," New Mexico Soc. Prof. Engrs., *J. Am. Water Works Assoc.,* **44,** 435–441 (May 1952).
12. McClure, G. M., Eiber, R. J., Hahn, G. T., Boulger, F. W., and Masubuchi, K., "Research on the Properties of Line Pipe," Summary Report Catalog No. 40/PR, 1962, American Gas Association, New York.
13. Verein. d. Grosskesselbesitzer, 1955/56 Review.
14. Verein. d. Grosskesselbesitzer, 1954/55 Review.
15. Texter, H. G., "Why Oil-Well Tubing and Casing Fail in Tension, and Why They Collapse," *Oil Gas J.,* **54,** 86–96 (July 4, 1955).
16. Kauczor, E., "Brüche durch unsachgemässes Schweissen," *Schweissen und Schneiden,* **15,** 430–434 (1963).
17. Soete, W., "Brittle Strength of Cracked Steels," *Rev. Soudure-Brussels,* **15,** 147–152 (March 1959).
18. Bier, P. J., "Anderson Ranch Dam Penstock Test Fracture Repaired by Welding," *Welding J.,* **32,** 313–319 (1953).
19. Banik, E., "Zerknall eines Druckluftbehälters in Folge schlechter Schweissung," *Schweissen und Schneiden,* **2,** 36–38 (1950).

———chapter 14

HEAT TREATMENT

PURPOSES

The primary purposes of heat treatments are to reduce residual stresses or to effect changes in the metallurgical structure.

Where the reduction of residual stresses is the primary consideration, an improvement in the resistance to failure may be desired in service environments involving the following:

(1) Fatigue loading (mechanical or thermal)
(2) Brittle fracture conditions (usually at low temperatures)
(3) Stress corrosion cracking
(4) Localized corrosion
(5) Geometrical stability

Heat treatments to produce desired metallurgical changes may be made to provide:

(1) Homogenization in the structure
(2) Solution of specific phases
(3) Stabilization of specific phases
(4) Age hardening
(5) Formation of specific structures

The results may be hardening, strengthening, softening, improved formability, better weldability or other properties.

Proper heat treatment is as essential to a soundly fabricated pressure vessel and piping system as are good design, forming, and welding procedures. Many service failures can be ascribed to careless or improper heat treatments, or to the elimination of heat treatment where it is, or should have been, called for.

In determining the proper heat treatment, consideration should be given to temperature, holding period, heating and cooling cycles, and the method of heating employed.

Ideally, materials for all components in a pressure vessel or piping system should have the same mechanical and metallurgical properties. However, the properties of a material can be varied substantially by fabricating and welding operations and by heat treatment. Although this is more generally recognized with regard to carbon and low alloy steel materials, significant variations in properties and service behavior can also be obtained by localized heat treatment of stainless steels and nonferrous metals.

Among the failures ascribed to improper heat treatments are:

(1) excessive hardness differences
(2) variations in metallurgical structures
(3) localized stresses
(4) surface defects
(5) high transition temperature
(6) reduced creep strength

Consideration of the proper heat treatment, and the consequences of failures when heat treatments have not been properly done, or have been eliminated, covers the whole thermal history of a pressure vessel or piping component. It includes preheat as well as postheat treatments.

PREHEAT TREATMENT

On many low-alloy steels, heating prior to welding is an effective means of avoiding weld or heat-affected zone cracking. Since the latter occurs underneath the weld bead, it is also referred to as underbead cracking. (See Chapter 12.)

An example of underbead cracking near the surface of a $\frac{1}{2}$Mo steel pipe is illustrated in Fig. 14-1.

The beneficial effects of preheat treatments are primarily the re-

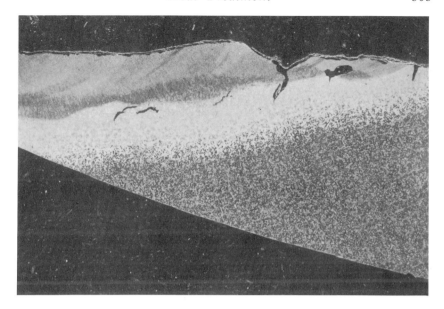

Fig. 14-1. Underbed cracking in ½ Mo steel pipe welded without preheat

duction of stresses and the increase in ductility in the weld heat-affected zone, making it more resistant to cold cracking. When hydrogen is present in the weld or heat-affected zone, there is a greater tendency toward cold cracking; this can be reduced with increasingly higher preheat temperatures of up to 400 to 600°F.

The need for proper preheat is generally greatest when welding the root pass in butt welds. It is also more important when welding processes are used that deposit relatively thin passes in the root, as is the case with inert-gas tungsten-arc welding. When heavier weld deposits are made in the root, as for example, by shielded metal-arc welding, there is less dependence on preheat treatments to avoid cracking.

Heavy wall sections of unfavorable weld joint contours may also necessitate preheat treatments, sometimes at temperatures higher than those normally recommended. Figure 14-2 illustrates cracking in the pipe base metal ⅜ in. away from, and parallel to, a heavy fillet weld. The piping material was ASTM A106 Grade C containing 0.31% carbon. The weld, representing heavy restraint conditions, was made without preheat. The pipe thickness was 1¾ in. and the fillet weld measured 1¼ in. high by ½ in. wide.

Fig. 14-2. Cracking adjacent to a heavy fillet weld made without preheat in carbon steel pipe containing 0.31 C.

Lack of preheat was considered the primary cause of another failure in a Mo steel pipe (0.19C, 0.3Mo, 0.6Mn and 0.3Cr) where cracking occurred in the longitudinal seam weld in a 28½-in. OD by ½-in. wall pipe. The lack of a preheat resulted in high hardness values in the heat-affected zone.[1]

In some materials, preheat treatments may also reduce susceptibility to hot cracking.

The interpass temperature during welding should, of course, be maintained at the levels of the preheat temperatures.

On chromium-molybdenum alloy steels up to the 3Cr–1Mo grades, the preheat (interpass) temperature should be maintained until at least ⅜ in. of weld metal is deposited, or until 25% of the welding groove is filled.[2] On 5Cr–½Mo alloy steels and other more highly alloyed grades, the preheat (interpass) temperature should not be interrupted until the weld is completed. Or, the partially completed weld can be immediately stress relieved at a minimum temperature of 1200°F for at least 30 minutes.[3]

Recommendations for preheat treatments are included in Table 14-1.

POSTHEAT TREATMENTS

Uniform Hardness

On carbon and alloy steels that have been heated to high temperatures, rapid cooling rates may significantly increase the strength and hardness in the heated area. This can be the result of actual heat treatments, or it may be caused by cutting, welding, hot forming, or flame straightening operations. Differences in chemical compositions or carbon content of the steel or weldment may also produce variations in the response to heat treatments.

Conditions for failure may occur in steels when significant hardness differences are not reduced by subsequent heat treatment, and when the service involves mechanical and/or thermal fatigue.

Cracking along a butt weld between a $\frac{1}{2}$Mo cast steel valve body and pipe in a main steam power plant piping system was illustrated in Fig. 8-14. The higher hardness in the weld heat-affected zone was not removed by a 1150°F postweld heat treatment. Had the heat treatment been done at 1325°F, the hardness values in the weld heat-affected zone would have been reduced considerably, and cracking would not have been likely.

Similarly, an insufficiently high temperature and too short a holding period were considered responsible for a brittle rupture in a $9\frac{1}{2}$-in. OD by 1.1-in. wall 1Cr–$\frac{1}{2}$Mo pipe. Cracking occurred along the fusion line between weld and heat-affected zone. Hardness values were substantially higher than normal.[4]

Recognizing these problems, a number of codes are now including hardness limitations in their requirements. For example, the ANSI Code for Pressure Piping, Section B31.3, *Petroleum Refinery Piping*, lists the following maximum permissible hardness values:

Alloy Groups	Maximum Hardness Brinell
$\frac{1}{2}$Cr–$\frac{1}{2}$Mo	225
1Cr–$\frac{1}{2}$Mo, $1\frac{1}{4}$Cr–$\frac{1}{2}$Mo, 2Cr–$\frac{1}{2}$Mo	225
$2\frac{1}{4}$Cr–1Mo, 5Cr–$\frac{1}{2}$Mo, 9Cr–1Mo	241
12Cr, 15Cr	241

In some low-alloy steels used in high temperature service, the elimination of significant hardness differences across weld joints may be difficult. For example, vanadium steels (0.10–0.18C, 0.3–0.6Mn, 0.50–0.65Mo, 0.25–0.35V) used in Europe in steam piping at about 1000°F tend to exhibit significantly higher hardness values in welds made from electrodes of corresponding compositions. Stress relieving for 2 hours at 1290°F leaves the hardness differences shown in Fig. 14-3.[5] In these materials, higher stress relieving temperatures, particularly in field erected joints, are not considered desirable due to their adverse effects on the creep-strength properties of the base metal. It remains to be seen if these differences will produce adverse results in $8\tfrac{5}{8}$-in. OD by 0.87-in. wall steam power piping operating at about 1000°F, unless thermal and mechanical fatigue conditions are of minor magnitude. The relatively light wall thickness involved may also reduce the potential tendency toward cracking.

Close Temperature Control

Uniform heat treatments and close temperature control are particularly important in some of the higher strength materials such as

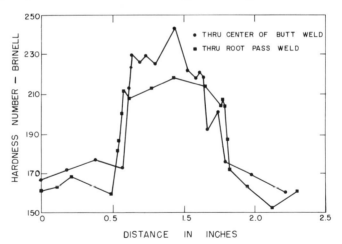

Fig. 14-3. Variation of hardness from base metal across weld deposit in Mo-V steel pipe (0.10-0.18 C, 0.15-0.35 Si, 0.30-0.60 Mn, 0.50-0.65 Mo, 0.25-0.35V) welded with Vanadium steel electrodes and stress relieved 2 hrs. at 1290F.[5] (Values originally reported in Vickers have been converted to Brinell)

Cr–Mo–V and Ni–Cr–Mo–V steels and alloys depending on solution and age hardening heat treatments.

For example, Cr–Mo–V steels heat treated at an austenitizing temperature of 1850°F following by tempering at 1225°F are particularly notch sensitive. Such materials have failed in steam turbines involving ruptures in rotor blade groove walls.[6] The high temperature notch sensitivity is substantially reduced by carefully controlled austenitizing at 1750°F followed by tempering at 1225°F.

Nonuniformity in heat treatment has also been considered responsible for residual stresses in 17-4 PH stainless steel control-rod index tubes in nuclear reactors. The failures were associated with stress corrosion cracking.[7] Uniform heat treatment at 1100°F minimized susceptibility to stress corrosion.

Solution Heat Treatment of Corrosion Resisting Stainless Steels

An example previously illustrated (Fig. 4-14) involved intergranular corrosion in certain types of austenitic stainless steels, such as Types 304, 309, 310 and 316. In the heat-affected zone adjacent to the weld where the steel is heated to between 900 and 1500°F, chromium carbides tend to precipitate along the grain boundaries of the metal. Some corrosive solutions attack these grades of stainless steel along the grain boundaries, as shown in Fig. 14-4. Although it is generally more convenient to specify stainless steel grades that are not susceptible to intergranular corrosion, there are instances where it is more desirable or less costly to perform heat treatments, which in effect return to the grain boundary the normal corrosion resistance of the stainless steel.

Fig. 14-4. Intergranular corrosion of Type 304 stainless steel in part of the weld heat-affected zone heated to between 900 and 1500°F during welding.

The so-called solution heat treatment generally is done in furnaces between 1900 and 2050°F for about ½ to 1 hour. On light wall thicknesses, as pipe Schedules 5 and 10, a few minutes at 2000°F may actually be sufficient. Ideally, this should be followed by rapid cooling, either by quenching in water or by a water spray. Although this may not present a problem with straight piping or tubing, more complicated fabricated piping assemblies may suffer excessive distortion. Air cooling may then be the only alternative, even though some reprecipitation of the chromium carbides may occur.

Solution Heat Treatment of Superheater Tubing

A number of unexpected service failures have occurred in titanium stabilized Type 321 stainless steel superheater tubing. The failures were preceded by swelling and were associated with creep and occurred after at least 40,000 hours of service.[8,9]

The primary cause was that the heat treatments normally used do not develop the potential creep strength levels in these stainless steel grades.

Superheater tubes are normally produced by cold working methods. Subsequent heat treatment was generally done at temperatures between 1600 and 1800°F. The resulting metallurgical grain structure was considered to be extremely fine-grained, having an ASTM grain size of about 11 or finer.

Heat treatment at 2000 to 2100°F produced a coarse grain size of ASTM size 7 or coarser. This is generally associated with the solution of titanium carbides (TiC). This resulted in a significant increase in the creep-rupture strength.

By raising the solution heat-treatment temperature from 1900 to 2050°F, the rupture strength at 1200°F is increased at 10,000 hours from about 13,000 to 17,000 psi, and at 100,000 hours from about 9,000 to 13,000 psi.[10]

Applicable ASTM Specifications now require that Type 321 tubing for high-temperature service, when produced by cold working, be heat-treated at a minimum temperature of 2000°F. As evidence of this heat treatment, a minimum grain size of No. 7 is required.

Stainless steel tubing grades for high-temperature service are now given an "H" designation such as Type 321H, 304H, or others.

Quenching and Tempering Heat Treatments

In a number of alloy steels superior properties are obtained by quenching from above 1600°F followed by tempering at about 1250 to 1300°F. In these steels, in the presence of carbide-forming elements, tempering produces initially very finely dispersed alloy-rich carbides. This results in a marked retardation of the softening process. Sometimes, even a slight increase in hardness may be obtained.[11]

The purpose of this heat treatment is to produce tempered martensite which exhibits a substantially greater toughness and better notch impact properties than a pearlitic structure of the same hardness.

These steels are most extensively used in pressure vessels. However, some use of these steels and heat treatments is also made in piping, particularly in Europe.

After quenching, prior to the tempering heat treatment, these steels may be susceptible to cracking, particularly along the surface. The cracks generally appear similar to transcrystalline stress corrosion but may also be intercrystalline, or both, Fig. 14-5. Even though subsequent tempering improves the toughness of the steel, the cracks reduce significantly the notch toughness and lower the transition temperature.

Service failures have been reported in $12\frac{1}{2}$-in. OD by 0.4-in. wall piping used in a steam heating system.[1] Photomicrographs of these failures normally show decarburization of the steel along the original cracks. The decarburization occurs during the subsequent tempering.

Indiscriminate Elimination of Heat Treatment

Just because codes permit the elimination of postheat treatments does not mean that they should be eliminated in every instance. For example, stress relieving is generally not required by codes for $1\frac{1}{4}$Cr–$\frac{1}{2}$Mo, and $2\frac{1}{4}$Cr–1Mo alloy steel piping that is 4 in. in diameter or smaller and has a wall thickness less than $\frac{1}{2}$ in.[12]

Fabrication operations can cause strains or stresses of significant magnitude in the pressure vessel, pipe, or tubing being fabricated. For example, residual stresses of the order of 30,000 to 40,000 psi

310   *Defects and Failures in Pressure Vessels and Piping*

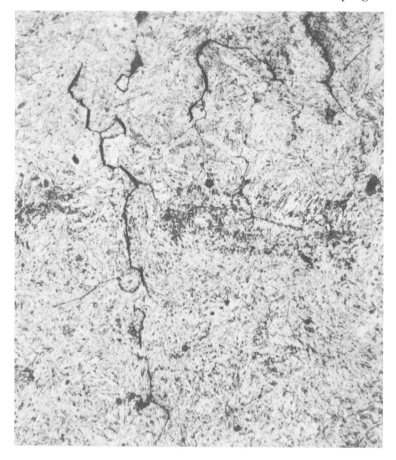

Fig. 14-5. Cracking in Cr-Mo-V steel surface caused by improper quenching prior to tempering. (Magnification 500×)

have been measured on the surfaces of cold bent superheater tubes and in weld zones of superheater headers.[13]

A case of failures occurring in $1\tfrac{1}{4}$Cr–$\tfrac{1}{2}$Mo superheater tubing illustrates that the operations involved should determine the advisability of heat treatments. The tubing had been cold bent and lugs were subsequently welded onto the outside. The tubing was then cleaned by pickling and finally erected into the boiler. The failures that occurred resulted from stress corrosion cracking in the cold bent area. The cracking started at the inside of the tubing, underneath

Heat Treatment

Fig. 14-6. Stress-corrosion cracks across wall of 1¼ Cr–½ Mo superheater tube in cold bent area underneath location where hanger lugs had been welded on tube section. The 2½-in. OD by 0.26-in. wall tube had not been subsequently stress relieved. (Magnification 8×)

the lugs, as illustrated in Fig. 14-6. Stress relieving of the tubing after the cold bending and welding would have avoided these failures.

Lack of postheat treatments have been found responsible for substantially increased rates involving pitting corrosion due to oxygen in a feedwater preheater. The pitting on the inside of the 0.25Cr–0.35Mo tubing was concentrated at the bent areas where the hardness had increased to 228 Brinell and in 1Cr–½Mo steel welds exhibiting hardness values between 228 and 270 Brinell.[14]

Service failures have occurred also in 16Cr–13Ni–Cb stainless steel superheater tubing which had been cold bent. For example, transcrystalling stress corrosion cracking has been reported in the cold bent area where the hardness varied between 219 and 238 Brinell, whereas the hardness outside the bent area was 162 to 171 Brinell. The maximum ovality in the bent section of the 1.26-in. outside diameter by 0.138-in. wall tubing was 6 to 7%.[15] Particularly welds in the stainless steel materials have been susceptible to corrosion fatigue cracking.[16, 17] As a result, stress relieving at temperatures of 1650 to 1920°F has been recommended for these materials with 1920°F representing the preferred temperature.[18]

A failure by corrosion in a carbon steel pipe that had been cold bent is shown in Fig. 14-7. The service was concentrated sulfuric acid. Corrosion occurred only in the outer arc of the pipe bend, where residual tensile stresses remained after cold bending. Attack did not occur along the neutral axis nor along the inside arc, where compressive stresses were present. The excessive tensile stresses along the center of the outer bending arc might even suggest the presence of a weld. Actually, however, the pipe shown in Fig. 14-7 is seamless. A heat treatment to eliminate the residual stresses after bending could probably have avoided this type of corrosive attack.

A number of cracking failures have occurred in carbon and alloy steel vessels and piping where repair welding was done on butt welds that had been previously stress relieved. In these cases, the repair welding was not followed by another stress relieving operation.[19]

Sometimes, in order to eliminate postheat treatments, low alloy steels are welded with austenitic stainless steel electrodes such as 25Cr–20Ni (Type 310) or 25Cr–12Ni (Type 309). When properly preheated to a minimum of 450°F, these joints may not crack, particularly when sections of relatively light wall thickness (less than ½ in.) are involved, so that postheat treatment may not be absolutely necessary. Stress relief heat treatments, however, represent still better practice to avoid high residual stresses and high hardness levels in the base metal. Figure 14-8 illustrates cracking in a vertical type platformer heat exchanger. A 12-in. OD by 0.330-in. wall pipe section of 5Cr–½Mo steel was welded with 25Cr–20Ni stainless steel electrodes into a floating head. A 10-in. long crack was found after 6 years service at 240°F at the toe of the weld around the pipe circumference. Improper preheat and lack of a postheat left residual stresses which made the steel susceptible to stress-corrosion cracking in an atmosphere of H_2S.[20] The degree of hardening (Vickers diamond hardness) is shown in Fig. 14-9.

Nonuniform and Localized Heating

Localized heating as a result of welding, hot forming, or postheat treatments may cause metallurgical changes of sufficient significance

Heat Treatment

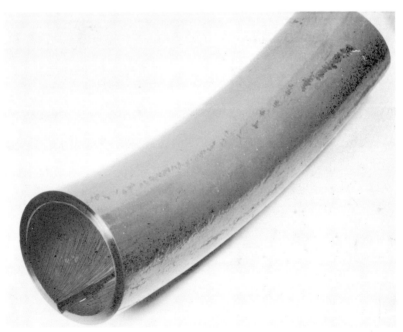

Fig. 14-7. Corrosion by concentrated sulfuric acid of seamless carbon steel pipe section which had been cold bent.

314   *Defects and Failures in Pressure Vessels and Piping*

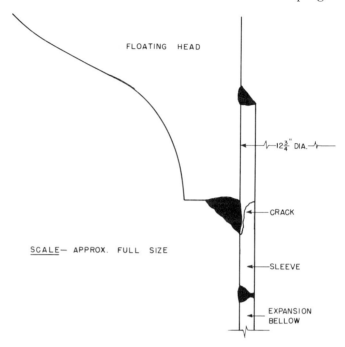

Fig. 14-8. Location of cracking in heat-affected zone of 5 Cr–½ Mo base metal adjacent to stainless steel weld deposit.[20]

Fig. 14-9. Cracking in vertical platformer heat exchanger associated with improper heat treatment. Vickers hardness numbers illustrate base metal hardening.[20]

Fig. 14-10. Cracking in 40-in. OD by 1.26-in. wall tubular heat exchanger caused by improper heat treatment.[21]

to lead to failures, particularly by corrosion. An example of this in carbon steel pipe in hot water service was illustrated in Figs. 11-1 and 11-2.

Nonuniform stress relieving by local methods may do more harm than good, and may, in effect, increase the level of stress.

Cracking in the 40-in. OD by 1.26-in. wall shell of a tubular heat exchanger in a platinum thermal reformer made of $1Cr-\frac{1}{2}Mo$ alloy steel is illustrated in Fig. 14-10. Failure occurred after 4 years of service at 950°F. The longitudinal crack originated along the weld around the reinforcing ring. Service conditions involving thermal fatigue caused crack propagation through the butt weld toward the welding neck flange. The crack permitted the leakage of gasoline causing fire.[21] Nonuniform heat treatment leaving substantial stresses was considered the primary cause.

Some fabricators who water quench fabricated pipe sections subsequently correct the distortion by flame straightening, which in-

316   *Defects and Failures in Pressure Vessels and Piping*

volves localized heating. The problems that may be encountered will be illustrated under the discussion of flame straightening.

Excessive Heating

Excessive heating or holding at certain temperatures may produce undesirable changes in the metallurgical structure.

Periodic annealing of riveted and welded gas cylinders in service for several decades resulted in substantial carbide spheroidization. The resulting loss of strength led to ruptures of the cylinders.[22]

FLAME STRAIGHTENING

Because of the normal expansion and contraction of metals, the welding of vessels, tanks, and piping assemblies tends to distort the original alignment of the various individual components. The degree

Fig. 14-11. Examples of distortion in two pipe headers. Severe distortion occurred in left header due to welding of nozzles on one side only. Very little distortion occurred in right header where nozzles were welded on opposite sides. Left header was straightened by localized flame heating, followed by final stress relieving.

of distortion depends on the metal or alloy welded; its size, shape, and thickness; the tacking and alignment; the welding process, procedure, and sequence; and the care taken by the welder. Distortion of vessels and pipe sections also increases when attachment or nozzle welds are made along one side only, as illustrated in Fig. 14-11.

The method most widely employed to correct misalignment in vessel and piping components is the alternate heating and cooling of areas adjacent to the welds. The consequent expansion and contraction of the metal then gradually draw the assemblies into proper alignment. On carbon steel, the cooling is normally done by applying water to the heated area. With water spray, the impingement of the water produces the most rapid cooling rate. More gradual flowing of water over the hot surface may form a steam blanket, reducing the cooling rate. On alloy steel materials, the application of water is generally not permitted since it may result in excessive localized hardness. Cracking may then result in environments involving thermal or mechanical fatigue. Air cooling, therefore, must suffice.

Flame straightening, even when not followed by water quenching, may leave residual stresses of substantial magnitude. In services where stress corrosion can occur, cracking may take place in the areas heated during the flame straightening.

Figure 14-12 illustrates cracking in the center of the area heated when a carbon steel pipe was flame straightened. The pipe carried 40% caustic at 240°F. The increase in hardness in the flame straightened area is shown graphically in Fig. 14-13.

Water quenching may also raise the transition temperature of some steels due to quench aging or martensite formation. This may be undesirable in applications where good notch toughness is important. However, the effect varies with the type of steel and the presence of alloying elements.[23] The increase in transition temperature is greatest about $\frac{1}{2}$ to 1 in. from the edge of the quenched zone.[24] Subsequent postheat treatments of 650°F [25] or at higher stress relieving or normalizing temperatures may again improve the notch toughness of many steel compositions.

Local heating for flame straightening of Types 304, 309, 310 and 316 stainless steel piping assemblies tends to cause intergranular carbide precipitation. In corrosive environments where these carbides are attacked failures may result.

Fig. 14-12. Cracking in center of area heated to accomplish flame straightening. Heated area left high residual stresses causing susceptibility to stress-corrosion cracking. Crack revealed by fluorescent penetrant examination.

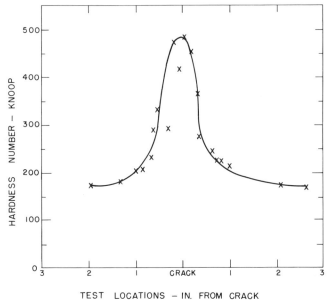

Fig. 14-13. Variation in hardness across carbon steel area locally heated for flame straightening.

Heat Treatment 319

Since stainless steel pipe fabrications are often cleaned by pickling, the scale evidence of locally heated areas may be removed. The personnel in the plant where the piping is then installed may not be aware of the sensitized condition of the stainless steel in the heated areas.

PROPER HEAT TREATMENT

In addition to proper temperatures and holding periods, heat treatment must employ proper rates of heating, and even more important, of cooling. The heating and cooling rates should be established on the bases of weld joint configuration, wall thickness, and the method of heating. More rapid rates of heating and cooling may be permissible on circumferential pipe butt welds than on angular nozzle welds or other welds between sections of dissimilar wall thicknesses. More rapid heating and cooling rates may also be permissible with induction heating, where the heating is accomplished within the pipe wall, than with surface heating methods, which depend on thermal conduction across the pipe wall.

Heat treatments and hot working temperatures recommended for carbon, alloy, and stainless steel piping materials are summarized in Table 14-1.

On vessels and pipe of larger diameters and heavier wall thicknesses, uniform heating around the pipe circumference is particularly desirable. In shop fabrication, this is most effectively accomplished by full furnace heating.

Induction heating, illustrated in Fig. 14-14, also insures uniformity of heat treatment. Relatively low frequencies of 60 to 400 cycles are generally used. In torch or electric resistance heating, the heat is slowly conducted from the point of application on the outside surface to the inside wall of the pipe. Induction heating generates heat essentially within the pipe wall. On pipe of heavier wall thicknesses, this has the advantage of providing more uniform heat throughout the wall and a smaller temperature difference between the outside and inside surfaces in the weld area. The lower the frequency, the deeper is the penetration of the heat induced into the pipe wall. While the differences between 60 and 400 cycle heating are insignificant for wall thicknesses up to about $1\frac{1}{2}$ to 2 in., the 60

Table 14-1. Recommended Temperatures for Preheat, Post Weld Heat Treatment, Hot Forming, and Heat Treatment after Hot Forming.

P Number	Material Group	Preheat Temperature, °F [1]	Post Weld Heat Treatment			Hot Forming Temperature °F	Heat Treatment After Hot Forming	
			Metal Temperature, °F	Holding Time [6]			Temperature, °F	Holding Time, Hr/in. Wall Thickness
				Hr/in. Wall Thickness	Minimum, hr			
P1	Carbon steel	not req'd [2]	1150-1250 [3]	1	½	1400-2000	May be necessary on heavy wall thickness or if heat for hot forming was uneven.	
P2	Wrought iron	not req'd [2]	—	—	—	1400-2000	—	—
P3	Maximum alloy: ¾% Cr; 2% total	300-600	1275-1350 [4,5]	1	1	1500-2000	1275-1350	1
P4	Maximum alloy: ¾-2% Cr; 2¾% total	400-600	1300-1400 [5]	1	1	1600-2000	1300-1400	1
P5	Maximum alloy: 2-3% Cr	400-600	1300-1450 [5]	1	1	1600-2000	1300-1400	1
	Maximum alloy: 10% total	500-600	1300-1425	2	2	1600-2000	A: 1550-1600 [7] N: 1650-1700 [7] T: 1300-1425 [7]	1 [10] 1 [11]

				Varies with alloy and application.		
P6	Martensitic stainless	500-700	1400-1500	1		1400-1500	1
P7	Ferritic stainless	not req'd	not req'd[8]	—	1500-1900	not req'd[8]	—
P8	Austenitic stainless	not req'd	not req'd[9]	—	1500-2100	not req'd[9]	—
P9	Nickel alloy	300-500	1200-1300	1	1600-2000	1200-1300	1

[1] Use lower temperature for sections up to ½-in. thick and higher temperature for sections over 2-in. thick, with intermediate temperatures for intermediate thicknesses.

[2] If the ambient temperature is less than 32°F, local preheating so that material is warm to the touch is required. Preheat may be desirable where the carbon content exceeds 0.30%, particularly in heavier wall thicknesses.

[3] Post weld heat treatment is not required for carbon steel if its thickness is less than ¾ in.

[4] Post weld heat treatment is required for ½ per cent molybdenum steel if its thickness is less than ½ in.

[5] Post weld heat treatment is not mandatory for socket welds and butt welds of joints 4 in. OD and less and wall thicknesses of less than ½ in. for all P3 and P4 groups and the 2¼ and 3 per cent Cr materials of the P5 group.

[6] The heating rate up to 600°F is optional; above 600°F, it is optional for light wall thicknesses. However, for pipe wall thicknesses over 1 in. for the P1-P3 groups, over ¾ in. for the P4-P5 groups (to 3 per cent Cr), and over ½ in. for the P5-P6 groups, the following heating and cooling rates above 600°F apply: For induction heating up to 1½-in. wall thickness, the maximum rate is 500°F per hr for both 60 and 400 cycles; for wall thicknesses of 1½ in. and above, the maximum rate is 500°F per hr for 60 cycles and 400°F per hr for 400 cycles. For furnace, gas, electric resistance, and other surface heating methods, the maximum rate is 400°F per hr, or 400°F per hr divided by one-half the wall thickness, whichever is smaller.

[7] Heat treatment after hot forming involves either annealing (A) or normalizing (N) and tempering (T) at the temperatures shown.

[8] Post weld heat treatment may be desirable for some grades to improve ductility.

[9] Annealing heat treatments may be necessary for some grades to improve corrosion resistance in specific environments.

[10] With furnace cooling at 50°F per hr to 1100°F.

[11] With air cooling.

322 *Defects and Failures in Pressure Vessels and Piping*

Fig. 14-14. Induction heating of heavy wall pipe insures uniformity of heat treatment.

cycle equipment provides more uniform and rapid heating on heavier wall thicknesses.

Torch heating of piping is done either with single burners or with ring burners. The latter is illustrated in Fig. 14-15. Heating with single burners should be limited to piping of smaller diameters (below 3 in.) and light wall thicknesses. Methods employing "softer" flames, such as propane or butane torch heating, are preferred over oxyacetylene torch heating for pipe heat treatment. Ring burners are also available for various gas mixtures, including propane, oxygen-propane, and oxyacetylene.

Single burner torches provide sufficiently uniform preheating to temperatures up to 600°F. For post weld heat treatment of circumferential pipe joints at stress relieving temperatures above 1000°F, however, precision-machined ring burners should be employed to insure uniform heating of the pipe around its entire circumference. At these temperatures, it is difficult to achieve uniform heating with single burner torches, and their use may result in metal upsetting, distortion and even cracking.

Fig. 14-15. Torch heating of pipe with ring burner should be done with specially made precision torches providing uniform heating around the pipe circumference.

If heavy attachment welds or large lug welds have been confined to a relatively small area on pipe sizes exceeding 5 in. in diameter and 1 in. in thickness, stress relieving should be done uniformly around the complete circumference. Localized stress relieving may be more damaging than beneficial.

REFERENCES

1. Verein. d. Grosskesselbesitzer, 1960/61 Review.
2. "Preheat and Postheat Welding Practices for Low-Chromium-Molybdenum Steel Pipe," Pipe Fabrication Institute Standard ES-8, Pittsburgh (Mar. 1964).
3. "Preheat and Postheat Welding Practices for Low-Chromium-Molybdenum Steel Pipe," Pipe Fabrication Institute Standard ES-12, Pittsburgh (Mar. 1964).
4. Verein. d. Grosskesselbesitzer, 1959/60 Review.
5. Wehrberger, F., "Erfahrungen beim Bau der Heizdampfleitungen aus dem Stahl 14MoV63 im Kraftwerk Nord des Volkswagenwerkes Wolfsburg," *Mitt. Ver. Grosskesselbesitzer*, **78**, 199–207 (1962).

6. Conrad, J. D. and Mochel, N. L., "Operating Experience With High-Temperature Steam-Turbine Rotors and Design Improvements in Rotor-Blade Fastening," *Trans. ASME*, **80**, 1210–1224 (1958).
7. Long, C. G., "Use of 17-4 PH Stainless Steel for Critical Reactor Components," *Nucl. Safety*, **3**, 39–41 (Sept. 1961).
8. Eberle, F. and Makris, J. S., "Fabrication and Annealing Factors Affecting Grain Size of 18% Cr − 8% Ni-Ti Superheater Materials in Steam Boilers," *Trans. ASME*, **82**, Series D, 855–866 (1960).
9. Leyda, W. E., "Investigation of the Effect of Heat Treatment on the Creep-Rupture Properties of the 18% Cr − 8% Ni-Ti Alloy," Babcock and Wilcox Company.
10. White, J. E. and Freeman, J. W., "Metallurgical Principles Governing the Creep-Rupture Strength of Type 321 ("18-8 + Ti") Austenitic Steel Superheating Tubing With Limited Extension to Type 304 ("18 Cr − 8 Ni") and Type 316 ("18 Cr − 8 Ni + Mo") Austenitic Steels," *Trans. ASME*, **85**, Series A, 119–146 (1963).
11. Grossmann, M. A. and Bain, E. C., *Principles of Heat Treatment*, 5th Ed., The American Society for Metals, Cleveland, 1964.
12. ASME Boiler and Pressure Vessel Code, Section I, Paragraph P-88, Notes 11 and 14; Section VIII, Table USC-56, Notes 11 and 13; ASA Code for Pressure Piping, Case 41; PFI Standard ES-8.
13. Wyatt, L. M. and Gemmill, M. G., "Experience With Power Generating Steam Plant and Its Bearing on Future Development," *Proceedings of the Joint International Conference on Creep* (1963).
14. Kaes, H., "Stillstandskorrosionen in Vorwärmern," *Mitt. Ver. Grosskesselbesitzer*, **82**, 62–64 (1963).
15. Verein. d. Grosskesselbesitzer, 1958/59 Review.
16. Noetzlin, G., "Das neue Kraftwerk Huls eine Anlage mit 300 at/600°C Frischdampfzustand," *Mitt. Ver. Grosskesselbesitzer*, **55**, 230–255 (1958).
17. Engl, A., "Der Einsatz austenitischer Stähle im neuen Kraftwerk Huls," *Mitt. Ver. Grosskesselbesitzer*, **55**, 255–264 (1958).
18. Ruttmann, W. and Brunzel, N., "10 Jahre austenitische Stähle im Kesselbetrieb," *Mitt. Ver. Grosskesselbesitzer*, **80**, 310–326 (1962).
19. Verein. d. Grosskesselbesitzer, 1957/58 Review.
20. Private Communication, Bataafse Internationale Petroleum Maatschappij N. V.
21. Private Communication, Mobil Oil Italiana.
22. Sopwith, C. D., "Particulars of Failures in Gas Cylinders," Admiralty Ship Welding Committee Document, No. FE 4-370, Dec. 1950.
23. Horikawa, K., Mimino, T., and Tomita, K., "Influence of Local Rapid Heating and Cooling Treatment Aiming at Strain Removal in Mild and High Strength Weld Structural Steel," *J. Iron Steel Inst. of Japan*, **42**, 330–331 (1956).
24. Yoshida, T., Matsunaga, W., Terai, K., and Murase, T., "Some Investigations of the Effect of Working Processes On High Tensile Steels Commercially Produced," *J. Japanese Welding Society*, **24**, 446–453 (1955).
25. Shank, M. E., *Control of Steel Construction to Avoid Brittle Failure*, Welding Research Council, New York (1957).

———chapter 15

CLEANING

Heating operations, such as hot bending or welding, or heat treatments, such as stress relieving, produce scaling (oxidation) on the inside and outside surfaces of the respective steel vessels and pipe sections. The thickness of the scale generally increases at higher temperatures and longer heating periods.

Rotary driven steel scrapers are normally employed to remove loose scale from carbon or low alloy steel pipe that has been hot bent. This method is called *turbining*. For many services, this is adequate. The thinner, more tightly adhering scale that remains is generally not considered harmful.

Some service conditions do require scale-free pipe. The cleaning methods normally employed are pickling and rinsing, shot blasting, and sand blasting. When improperly done, each of these methods can result in damage to the piping and, in some instances, can lead to service failures.

PICKLING

On carbon and low alloy steel pipe, complete removal of scale along the inside pipe surface by pickling is not always readily achieved. In many areas, the scale is loosened. In other areas where the scale does not come off readily, pitting may occur, as shown in Fig. 15-1. Excessive pitting or etching resulting from prolonged im-

326 *Defects and Failures in Pressure Vessels and Piping*

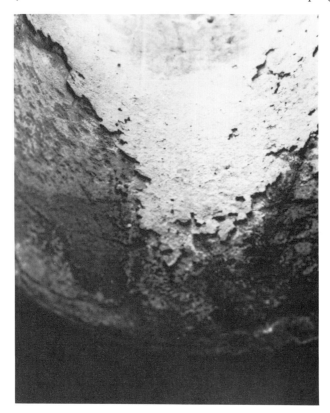

Fig. 15-1. Pitting in carbon steel pipe cleaned by acid pickling.

mersion in the pickling acid to remove all traces of scale may reduce the wall thickness of the pipe or tube below its acceptable minimum.

On some stainless steels, excessive pickling may cause substantial intergranular corrosive attack and loosen the grains of the metal, as illustrated in Fig. 15-2. As a result, a very fine metal powder is left on the inside of the vessel or pipe, which cannot be completely removed. This condition may be objectionable in cryogenic and other services requiring surgically clean piping.

Overpickling was considered responsible for failure of an outlet nozzle in a nuclear reactor involving Types 304 and 304L stainless steel materials. Overpickling of the outlet nozzle section appeared to damage the wall to a depth of as much as 0.040-in. Initiation of

Cleaning

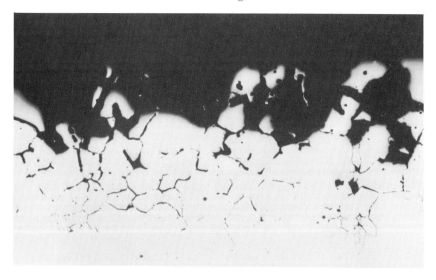

Fig. 15-2. Effect of overpickling along inside surface of Type 304 stainless steel pipe. The stainless steel pipe section has been properly heat treated and does not contain a significant degree of intergranular carbides. (Magnification 100×)

cracking was then caused by a field weld made about 1 in. from the damaged area. The failure caused leakage of heavy water.[1]

Removal of the pickling solution by alkaline washing and/or rinsing is an important step in the cleaning operation. Figure 15-3 illustrates stress corrosion cracking starting in boiler tubing. Apparently, acid traces had been retained in draw die grooves, probably for several months, in the presence of moisture. Had the tube been placed into service soon after pickling, the acid traces probably would have been washed out before doing significant harm.

Shot and Sand Blasting

An excellent and widely used method of cleaning carbon and low alloy steel piping is shot blasting. However, the shot blasting applicator, which sprays steel shot in an advancing rotary motion onto the pipe surface, should not be permitted to linger excessively at a particular location. An example of severe localized abrasion as a result of shot blasting is illustrated in Fig. 15-4. Shot blasting in the pipe was concentrated onto a point at the pipe inside. A cavity the

328   *Defects and Failures in Pressure Vessels and Piping*

Fig. 15-3. Stress-corrosion cracking starting from sharp corners in draw-die grooves of $1\frac{1}{4}$ Cr–$\frac{1}{2}$ Mo boiler tubing. (Magnification 250×)

size of a fist was produced. The wall thickness at this point was reduced from $1\frac{1}{2}$ in. to $\frac{1}{4}$ in. The pipe mill even had the pipe quality and uniformity certified by ultrasonic inspection.

TURBINING

Turbining of tube materials of lighter wall thickness, when not carefully done, has on occasion caused excessive thinning. This cleaning method is extensively done in maintenance of boiler, still and other tubing to remove scale and chemical deposits on the inside. Thinning, usually as a result of holding the turbining tube too long

Cleaning

Fig. 15-4. A 1¼-in. deep cavity at inside of 16-in. OD by 1½-in. wall steel pipe produced by improper shot blasting and certified by producer to be sound by ultrasonic examination.

in one position, may lead to subsequent service failures, often preceded by localized bulging.[2]

GRINDING OF WELDS

Increasing overemphasis is being directed to the "smooth" grinding of weld deposits. When vessel or pipe welds are to be radiographed, excessive roughness should be removed by grinding; however, there is rarely a need for completely smooth grinding. In fact, the overemphasis on grinding has frequently resulted in thinning of the vessel or pipe below the minimum acceptable wall thicknesses, particularly in heat-affected zones. Figure 15-5 illustrates cracking in a pipe after grinding had reduced the wall thickness 20% below the required minimum.

Fig. 15-5. Cracking across zone adjacent to weld where excessive grinding had reduced the wall thickness by 20%. The service involved high-temperature high-pressure steam in a power plant.

Mismatch at weld ends is particularly susceptible to excessive thinning as a result of grinding. Failure in a $3\frac{1}{2}$-in. OD by 0.157-in. wall 1Cr–$\frac{1}{2}$Mo superheater tubing occurred after 5000 hours at 985°F in the weld between a tube and flange. The inside diameter of the flange was somewhat greater than that of the tube end. The weld inside was ground to remove the notch effect of the weld. By smooth blending the surfaces, the wall thickness was reduced to 0.104 in.[3]

SHOT CLEANING OF BOILERS

Shot cleaning has been done on the inside of boilers to remove scale deposits which have formed during service.[4] This has had adverse results on the corrosion behavior of feedwater preheater tubing. The increased tensile stresses on the outside of the tubing caused by the impact of the shot particles has produced corresponding tensile stresses on the inside of the light wall tubing (approximately $\frac{1}{8}$-in wall) amounting to 8500 to 17,000 psi. The resulting corrosion occurring only in tubing containing water is associated with high

Cleaning 331

oxygen concentrations. The attack involves pitting corrosion and intergranular cracking with significant oxidation present in the cracks.[5] This latter attack, appearing similar to some types of stress corrosion cracking, has also been described as *oxygen stress corrosion*. (See Chapter 17.)

IMPROPER CLEANING FOR REPAIR WELDING

Where sections or components and pressure vessels and piping systems are to be replaced, proper precautions must be taken to insure that the old components do not contain surface or intergranular contaminations. Scale, deposits of copper, are frequent causes of trouble when not removed from the weld area.

Figure 15-6 illustrates intergranular cracking caused by penetration of copper into a boiler tube which had contained copper deposits along the inside surface prior to welding. The high heat of welding caused the copper to penetrate into the heat-affected zone of the steel along the grain boundaries.[6]

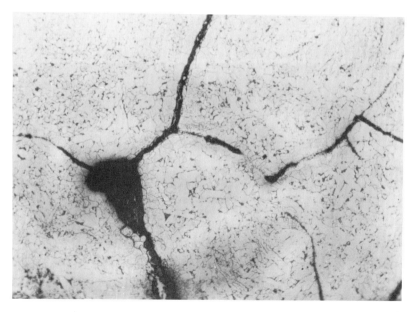

Fig. 15-6. Intergranular penetration of copper into steel tube caused by improper cleaning of copper deposits prior to repair welding.[6] (Magnification 250×)

REFERENCES

1. Gall, W. R., "Heavy-Water Leaks at Savannah River," *Nucl. Safety*, **3**, 87–90 (Sept. 1961).
2. Wright, E. C. and Habart, H., "Typical Failures of Still Tubes in Refineries," *Metal Progr.*, **34**, 573–578 (1938).
3. Verein. d. Grosskesselbesitzer, 1958/59 Review.
4. Chappell, R. E. and Meyer, R. D., "Shot Cleaning," *Combustion*, **30**, 39–45 (Dec. 1958).
5. Jahn, E. and Deeskow, M., "Rohrschäden unter gleichzeitigem Einfluss von Spannung und Sauerstoff," *Mitt. Ver. Grosskesselbesitzer*, **79**, 294–300 (1962).
6. Private Communication, Bataafse Internationale Petroleum Maatschappij N. V.

————chapter 16

SERVICE FAILURES—STRESS AND FATIGUE

CAUSES OF FAILURES

Pressure vessels, tanks, and piping which have been properly designed, assembled of sound materials, carefully and responsibly fabricated and thoroughly inspected normally should not be expected to fail in service. However, in some instances, the service conditions or environments are more severe or destructive than had been anticipated in the original design and when the materials had been selected.

Thus, the failures considered in this Chapter are those where the service is more severe than the pressure vessel, tank or piping system is capable of withstanding. Singly or in combination, many specific causes can lead to these failures. Some of the more common ones include:

(1) Excessive stresses
(2) Overpressure
(3) External loading
(4) Mechanical fatigue or shock
(5) Thermal fatigue or shock
(6) Overheating
(7) Hydrogen embrittlement
(8) Hydrogen blistering

334   *Defects and Failures in Pressure Vessels and Piping*

In some instances, the vessel or pipe may have been used on purpose at higher pressures or temperatures than had been the initial intent. Sometimes, this may be done inadvertently when pressure or temperature gages provide incorrect information.

Failures have also occurred when malfunctioning of control equipment or errors by operating personnel upset normal operational temperature and pressure balances.

Finally, many service environments involve fatigue conditions which often cannot be completely anticipated in the design, or which

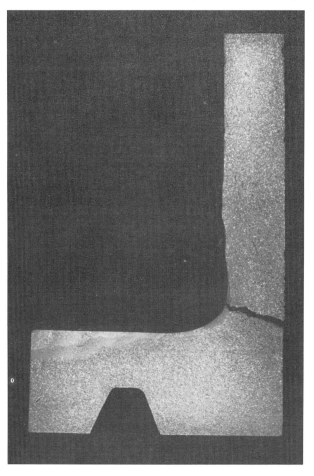

Fig. 16-1. Wear and cracking in lap joint pipe caused by uneven tightening of flange.

Service Failures—Stress and Fatigue 335

are more severe than the engineering materials now in use can withstand over long periods of time.

EXCESSIVE STRESSES

Although good design will normally avoid excessive stresses, improper assembly or supporting can result in excessive stresses leading to service failures.

Figure 16-1 illustrates cracking around $\frac{1}{3}$ of the circumference of a $\frac{1}{2}$Mo steel lap joint flange. Failure of the 8-in. diameter flange in service at 850°F was primarily due to uneven tightening of the flange ring. This resulted in tensile stresses. Thermal fatigue involved in the service also caused wear between the flange ring and the pipe wall near the lap, reducing the wall thickness.

OVERPRESSURE

Failures caused by overpressure have occurred in some chemical process equipment where process reactions did not proceed according to plan resulting in a severe pressure rise. Although normally such equipment is protected by safety valves, on occasion they have not functioned properly, leading to rupture of the equipment.

Fairly common are failures caused by freezing of water or other liquids in piping systems. An example of such a failure in a 2½-in. OD by 0.195-in. wall steel pipe used in a dry sprinkler system is illustrated in Fig. 16-2. Water had inadvertently flowed into the system and had not been removed.

Hydrostatic pressure testing can also occasionally result in overpressures and lead to rupture of the vessel or piping being tested.

Fig. 16-2. Overpressure failure caused by freezing of water in piping system.

EXTERNAL LOADING

External loading of vessels, tanks, and piping may be caused by supports, legs, brackets, hangers, and lugs. Walkways, platforms, and ladders may also transmit external loads.

External loading may be caused by the movement or settling of foundations. Improperly sized openings through concrete walls may also cause external loadings on vessels and piping when normal thermal expansion of the piping connected to vessels is restrained.

Occasionally, wind and ice become the cause of external loadings. Although rarely the primary cause of failure, these conditions may contribute to other factors already initiating or propagating cracking.

MECHANICAL FATIGUE AND SHOCK

Definitions

Fatigue represents the conditions leading to fracture under repeated or fluctuating stresses having a maximum value less than the tensile strength of the material. Fatigue fractures are progressive, starting as minute cracks that grow under the action of a fluctuating stress.

In pressure vessels and piping, mechanical fatigue can be caused singly or in combination by the following conditions:

(1) Pressure cycling
(2) Variations in flow
(3) System changes
(4) External factors

Pressure cycling involves changes in the internal pressure. *Variations in flow* may represent nonuniform steam flow passing through a pressure-reducing valve, water hammer, steam hammer, water-steam cavitation, etc. *System changes* may be caused on a header or a pressure vessel by different expansions of the piping components connected to it. *External factors* represent vibrations produced by specific components such as may be caused by a compressor, valve or pump, or produced by varying air pressure or by varying flow pressure on underwater piping.

Service Failures—Stress and Fatigue 337

The *fatigue limit* is the maximum stress below which a material can endure an infinite number of stress cycles.

The *fatigue strength* of a material is the maximum stress that can be sustained for a specific number of cycles without failure, the stress being completely reversed within each cycle (unless otherwise stated).

The number of fatigue cycles leading to failure may vary from many million cycles to less than one. Failure in less than one cycle, or a few cycles, is often considered as *mechanical shock*.

Fatigue failures occurring in relatively few cycles (arbitrarily considered less than 10,000 cycles) are also described as *low-cycle fatigue*.

In *low-cycle fatigue*,[1-3] the magnitude and the range of the load usually are sufficiently large to cause plastic deformation in the material. The corresponding hysteresis in the stress-strain behavior may change from one cycle to the next. The damage leading to failure involves the decrease in available ductility due to strain hardening and the formation and growth of cracks.[4]

In *low-cycle fatigue testing*, changes occur continuously in the geometry and the material of the specimen. This complicates evaluation of the stress distribution and strains during the test. Consequently, test results are generally presented in terms of nominal stress or strain, based on the initial conditions of the test. Thus, the numerical values shown are relative.

MECHANICAL FATIGUE

Vibrations are present in many pressure vessels and piping systems. Generally, the level is sufficiently low not to cause cracking or crack propagation.

A failure producing a leak was reported in a gas compressor station during service as a result of fatigue.[5] The crack developed at a tack weld on the bottom of the header where the pipe was resting on a support. Even under continuous high temperature conditions, some mechanical fatigue failures have occurred, as shown in Fig. 16-3. Usually, these can be related to pipe movement, vibration in turbine casings, restraints preventing free movement, or similar conditions. Such failures often include a series of parallel cracks on one side of the pipe wall, as in Fig. 16-4.

Severe vibrations were the primary cause of failure in a 23-ft high

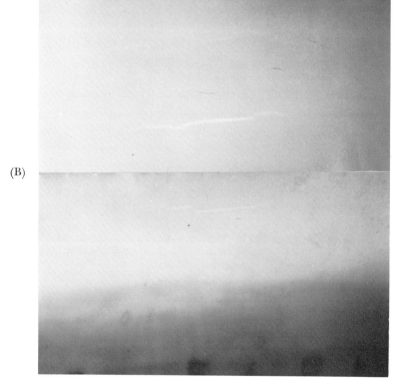

Fig. 16-3. Cracking in pipe section experiencing some mechanical fatigue. [(A) Photograph of surface. (B) Print of radiograph.]

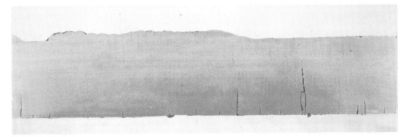

Fig. 16-4. Family of parallel cracks resulting from mechanical fatigue.

by 40-in. diameter carbon monoxide regenerator. The material was of a Type 321 stainless steel composition, approximately $\frac{1}{4}$ in. thick. The operating temperature was 1110 to 1290°F. The frequency of the vibration was in the audible range. After an estimated several hundred million cycles occurring in 17 days, cracking took place in the upper conical section at the location where holding lugs acted as fixed points, and in the supporting gussets. Since the welds were sound, the severe mechanical fatigue was the principal cause of the failures. In new equipment, such failures were avoided by the reduction of the vibrations.[6]

Obviously, notch conditions should be avoided or minimized in services involving mechanical fatigue.

MECHANICAL SHOCK

A few failures have resulted from severe mechanical shock. Since these failures frequently occur suddenly, they may have disastrous consequences.

Figure 16-5 illustrates a sudden failure of a cast iron flange. Failure was caused when a steam valve was opened. The steam passing into the piping hit a slug of water, which was thrown against the flanged elbow causing an instantaneous rupture. Movement of the pipe resulted in the loss of one life.

In June 1950, errors in operation and malfunctioning of equipment resulted in water hammer surges at the Oigawa Power Station in Japan. The resulting severe mechanical shocks burst a 9-ft diameter penstock causing the death of three workmen.[7]

Another example of a mechanical shock-type failure was caused

340 *Defects and Failures in Pressure Vessels and Piping*

Fig. 16-5. Mechanical shock failure in cast iron flange.

by sudden internal combustion produced by suction of air into a 26-in. diameter gas pipe in a compressor station.[5] The pipe fractured at several points involving both shear and cleavage-type fractures.

Severe shock loading during hydrostatic testing was considered the primary cause of failure of a 22-ft 6-in. diameter, 93-ft 9-in. long cylindrical boiler.[8] The failure occurred in 1963 at a British nuclear power station. The steel was a Mn–Cr–Mo–V material $2\frac{1}{4}$ in. thick.

THERMAL FATIGUE AND SHOCK

Definitions

Thermal fatigue and *thermal shock* are terms used to denote the effects on material life of temperature changes or alternating exposures at higher and lower temperatures. The differences between thermal fatigue and thermal shock are primarily related to rate change in temperature and the severity of the temperature gradient. Thus, when the service life is primarily determined by the number of thermal cycles, failure is generally ascribed to thermal fatigue.

Service Failures—Stress and Fatigue 341

However, when the severity of the temperature gradient or the rate of change in temperature is the primary cause of failure, such failure should be ascribed to thermal shock.[9]

Whether failure occurs by thermal fatigue or shock will vary with the material. The same temperature cycle may have a thermal fatigue effect on a ductile material and a thermal shock effect on a brittle material.

Pressure vessels and piping materials at the service conditions ordinarily encountered in steam power plants, refineries and chemical plants generally are not subject to thermal shock conditions. However, during fabrication of hardenable steels involving temperatures above 1600°F thermal shock conditions may lead to failure.

Thermal fatigue and shock may involve sudden heating (*heat shock*) or sudden cooling (*cold shock*).

Localized sudden heating will have the immediate result of producing compressive stresses in the heated area which may result in permanent upsetting. Subsequently, when the temperatures even out, residual tensile stresses may remain. If of sufficient magnitude, the stresses may enhance failure by creep usually involving intergranular cracking, or by further thermal fatigue or corrosion fatigue which may produce transcrystalline crack propagation.

Localized sudden cooling or quenching will immediately produce tensile stresses and may also result in permanent strain. Transcrystalline thermal fatigue cracking may occur, particularly if such cycles are applied repeatedly. Subsequent evening out of the temperature will leave compressive stresses in the quenched area, which are usually considered harmless.

Intermittent Operation

For the best metallurgical conditions, high temperature piping systems should be maintained continuously at the operating temperature.

Operating economics often make it advisable, however, to shutdown certain plant operations during specific periods. Steam power plants often have lower demand requirements at night and particularly on weekends. The reduced load requirements necessitate the shutdown of some boilers at night and a larger number of them on weekends. During a weekend shutdown, the piping carrying the

342 *Defects and Failures in Pressure Vessels and Piping*

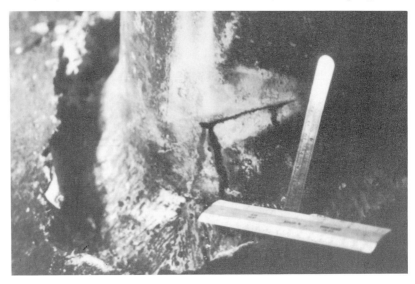

Fig. 16-6. Cracking in steam piping caused by thermal fatigue.

steam from a boiler to a turbine may cool from the 900 to 1100°F normal operating temperature to about 500 to 600°F. This causes contraction of the piping system, the boiler tubing, welded attachments, etc., in accordance with the coefficients of expansion of the materials involved. Sunday nights, when the power plant unit is returned to normal operation, the boiler tubing, steam piping, and turbine casings are heated up again by the steam produced in the boiler. Thermal expansion occurs in all metal components, and the piping systems expand to their normal positions. Over a period of months, service conditions involving thermal fatigue may be more harmful to materials, particularly at mechanical or metallurgical notches, than would be continuous operation at a specific high temperature over a period of many years.

Figure 16-6 illustrates cracking in a main steam lead caused by thermal fatigue. Sometimes, single cracks occur. More often, however, a family of parallel cracks can be found. The latter is particularly the case when thermal fatigue is caused by condensate carry-over. The cracks usually start on the inside of the pipe, since in most service environments, the inside surface is subject to more severe thermal shock than the outside surface. As shown in Fig. 16-7, these parallel

Service Failures—Stress and Fatigue 343

Fig. 16-7. Thermal fatigue cracking in steam piping caused by condensate introduced periodically into system and gradually propagating across wall thickness.

cracks gradually propagate until they reach the outside surface. Similar failures in main steam piping caused by condensate carry-over have been previously reported.[10, 11, 44] For example, on a ship a 5½-in. OD by ⅜-in. wall carbon steel pipe carrying superheated steam at 765°F failed after one year due to condensate periodically carrying over into the steam line from the small diameter combustion control piping.[11] A similar failure has been reported in 18-in. diameter by 0.937-in. wall 2¼Cr–1Mo alloy steel hot reheat piping after 30 months of service. Condensate through a pressure tap connection caused cracking in an elliptical area about 8 in. along the major axis and 4 in. along the minor axis.[12]

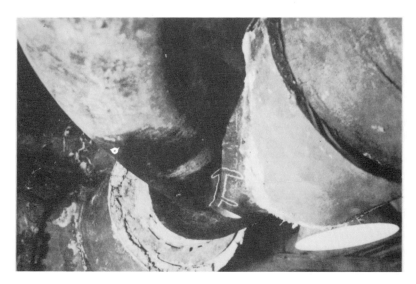

Fig. 16-8. Cracking in Y-connection caused by thermal fatigue.

344   Defects and Failures in Pressure Vessels and Piping

Fig. 16-9. Cracking through ½ Mo steel turbine casing caused by thermal fatigue.

Branch connections tend to exhibit a greater susceptibility to failure than straight piping. Figure 16-8 shows cracking in a Y connection where the main crack occurred along the edge of the weld with the perpendicular branches penetrating into the weld. Failure of the 12¾-in. OD by 1.125-in. wall ½Mo steel main steam pipe occurred after 15 years in service at 875°F. The intermittent operation involved five shutdowns each week.

Turbine casings, because of the varying thickness and design notches have become increasingly susceptible to thermal fatigue cracking as operating steam temperatures rose above 600°F.[13, 14] Figure 16-9 illustrates cracking through a ½Mo steel turbine casing. In these failures, the thermal gradients are produced either during

Service Failures—Stress and Fatigue 345

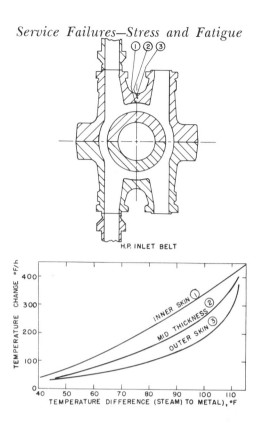

Fig. 16-10. Temperature conditions in turbine casing of high temperature inlet belt during start up.[15]

start-up of a cold turbine by the admission of hot steam onto a relatively cold metal surface, or during start-up of a hot turbine by the chilling of the hot metal surface by steam.

Temperature conditions in a turbine casing of the high-pressure inlet during start-up are illustrated in Fig. 16-10.[15] The most effective methods for reducing turbine shell cracking are improved operating procedures and the application of techniques that control shell-temperature differentials.[13]

Thermal fatigue failures similar to the cracking shown in Figs. 16-8 and 16-9 have occurred, also, in valves, fittings and other components.

Thermal fatigue failures in reducing valves caused by the sudden opening of the valves [16] have been avoided by maintaining the valve

346 *Defects and Failures in Pressure Vessels and Piping*

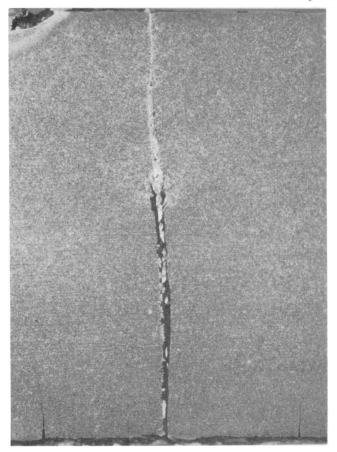

Fig. 16-11. Cracking across pipe wall in pressure reducing section where water is introduced. (Magnification 6×)

bodies heated with separately supplied high temperature steam and improved feed water quality.[17]

Operations Involving Temperature Cycling

Many failures have occurred in desuperheater sections involving water sprays on the inside of the pipe. The quenching effect, resulting in temperature cycling over a range of 300 to 500°F, causes failures across the pipe wall, as shown in Fig. 16-11. The insertion of stainless steel liner plates generally does not alleviate this condition.

Service Failures—Stress and Fatigue 347

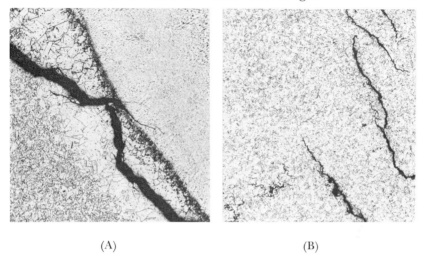

(A)   (B)

Fig. 16-12. Cracking of stainless steel liner section near weld (A) and in liner itself (B) in pressure reducing section. (Magnifications 60×)

As shown in Fig. 16-12, the liners tend to fail also, and usually just as readily as the pipe failed prior to the use of the liners.

A survey of experiences with desuperheaters conducted in 1958 in Germany involving different types of desuperheaters showed the following results.[18]

	No. of Desuperheaters in			
	Piping	Boilers	Total	%
No damage reported	212	128	340	72.5
Slight damage	14	15	29	6.2
Substantial Damage	75	25	100	21.3

The following factors were considered responsible for the conditions leading to the substantial damage.[18]

(1) Static loads caused by the internal pressure.

(2) Thermal fatigue caused by the differences in temperature between the steam entering and leaving the section and the *cold shock* effect of the cooling water coming in contact with the hot metal surface.

(3) Dynamic loads caused by vibrations due to the injected cooling

water and also caused in the piping by the vibrations produced by the steam passing through the pressure reducing valve.

Although most failures reported involved cracking across the wall, in one instance a severe rupture occurred. Some failures were also associated with base metal and weld defects.[18]

Another steam header failure caused by the periodic admission of desuperheater water is illustrated in Fig. 16-13. The header, $14\frac{1}{2}$ in. OD by $1\frac{1}{4}$ in. wall, is made of 1Cr–$\frac{1}{2}$Mo steel. Failure occurred after 100,000 hours at 950°F near the $3\frac{1}{2}$-in. diameter desuperheater outlet.[19]

A similar failure has been reported in the recirculating piping of a nuclear reactor containing water at 500 to 550°F. The 10-in. OD by 0.65-in. wall Type 304 stainless steel pipe developed a $2\frac{1}{2}$-in. long circumferential crack through the pipe wall. On the inside, numerous small cracks similar to Fig. 16-13 were located near the 1-in. diameter feed water inlet on top of the pipe. It was estimated that at least 1000 times during a five-year operating period, feed water at 70 to 130°F was introduced. Mechanical stresses were also

Fig. 16-13. "Dried Riverbed" type of cracking due to thermal fatigue in desuperheater outlet section.[19]

Service Failures—Stress and Fatigue 349

considered present every time the reactor was heated up due to the pipe being restrained by a 3-ft thick concrete wall.[20]

A thermal fatigue failure in Type 316 stainless steel tubing is illustrated in Fig. 16-14. The chemical plant service involved alter-

Fig. 16-14. Thermal fatigue failure in Type 316 stainless steel cooling coil due to frequent rapid heating to and cooling from 525°F. (A) photograph illustrating crack location (B) X-ray radiograph showing crack.

350   *Defects and Failures in Pressure Vessels and Piping*

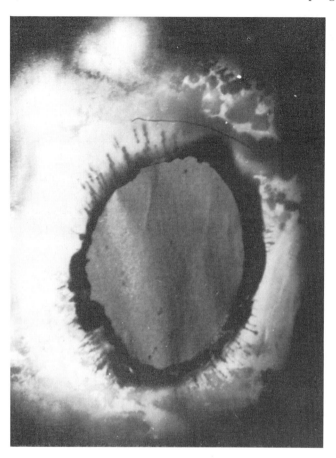

Fig. 16-15. Thermal fatigue cracking in stainless clad steel in paper mill on inside of steam inlet in sawdust cooker caused by cold liquor introduced in vessel directly above. Cracking delineated by liquid penetrant examination.

nate heating and cooling with the peak temperature being 525°F. Had thermal fatigue been anticipated, the weld joint should have been designed to provide a considerably more gradual change in surface contour.

Thermal Fatigue In Process Equipment

Figure 16-15 shows cracking due to thermal fatigue on the inside of a vessel at the steam inlet. Steam at a temperature of 550°F was

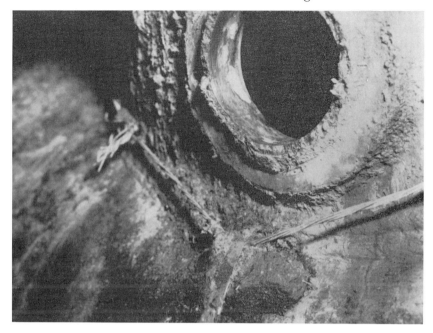

Fig. 16-16. Brittle failure across stainless steel pipe at fillet weld location was preceded by severe thermal fatigue cracking on inside of vessel.

brought into the vessel at this point. The "star" cracking, made evident by liquid penetrant inspection, is typical of thermal fatigue caused in this case by the introduction of a cold liquid above the steam inlet. After a few days of operation, some of the cracks propagated across the pipe inlet. A sudden rupture of the pipe, Fig. 16-16, was the result.

Notch Effects

The susceptibility to failure by thermal fatigue is usually increased by the presence of mechanical or metallurgical notches. When it is known or suspected that the service involves thermal fatigue, it is particularly important to design the system to avoid notch conditions. A number of examples of this are presented in Chapter 3 on Design.

Fabrication, welding and heat treating operations must be controlled to avoid or minimize the introduction of mechanical or metallurgical notches into a piping system.

352 Defects and Failures in Pressure Vessels and Piping

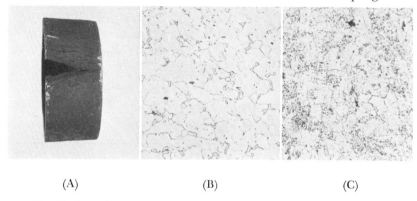

(A) (B) (C)

Fig. 16-17. Failure caused by overheating of carbon steel superheater tube (A), metallurgical microstructures on fireside (B) and opposite side (C). [Magnifications (B) and (C) 350×]

OVERHEATING

A substantial number of service failures occur from overheating, normally accompanied by creep. They have been most prevalent in steam power plant superheater tubing [21, 22] and in refinery cracking still tubing.[23]

Failures caused by overheating carbon steel and $1\frac{1}{4}$Cr–$\frac{1}{2}$Mo superheater tubing are illustrated in Figs. 16-17 and 16-18. Such fail-

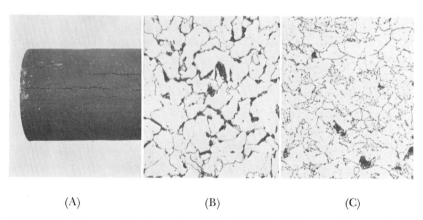

(A) (B) (C)

Fig. 16-18. Heavy oxidation and cracking in 1 Cr–$\frac{1}{2}$ Mo superheater tube (A), metallurgical microstructures on fireside (B) and opposite side (C). [Magnifications (B) and (C) 300×]

ures usually can be readily identified by considerable differences in microstructure.

When overheating results in gradual creeping, the cracking that follows in time tends to propagate along mechanical and metallurgical notches. Figure 16-19 illustrates a cracking failure in a 1Cr–$\frac{1}{2}$Mo steel vessel along the edge of the weld heat-affected zone. The change in microstructure in this zone also resulted in a hardness variation of approximately 20 points Brinell (converted from Knoop micro-hardness determinations).

Although most failures due to overheating are caused by external overheating, such failures have occurred also as a result of internal overheating.

A pressurizer vessel in a nuclear reactor failed as a result of internal overheating from electric heaters normally submerged in water. The vessel was 2 ft. 9 in. ID by 16 ft. 8 in. high of ASTM A264, Grade 3 (0.04 max. carbon) steel with Type 304L stainless steel fittings and internal cladding. The backing plate was ASTM A212, Grade B steel. The wall thickness was 2.95 in. including the $\frac{1}{8}$ in. cladding thickness. Failure involved a $\frac{3}{8}$-in. wide by $2\frac{1}{4}$-in. long rupture in the central girth seam weld. Normal immersion of the heaters in the water involved temperatures below 680°F and pressures of 2500 psi. Incorrect water level indication resulted in water levels being 20% lower than required, thus placing the two heaters into the steam zone. Metallographic examination indicated that temperatures approaching 1000°F were reached in the upper portion of the vessel.[24]

If service temperatures and pressures are such that some creep may be involved, the previous heat treatments may become a major factor in the life of the piping material.

A recent failure in $1\frac{1}{4}$Cr–$\frac{1}{2}$Mo superheater tubing designed for operation at 950°F but actually operated at 1040°F has resulted in a substantial reduction in the allowable stress values for this material at 1000 and 1050°F as follows:

Temp., °F	Old Stress Value, psi	New Stress Value, psi
1050	5500	4050
1000	7800	6550

354 Defects and Failures in Pressure Vessels and Piping

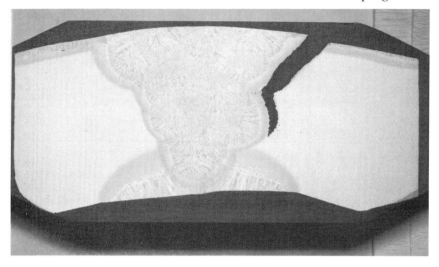

(A)

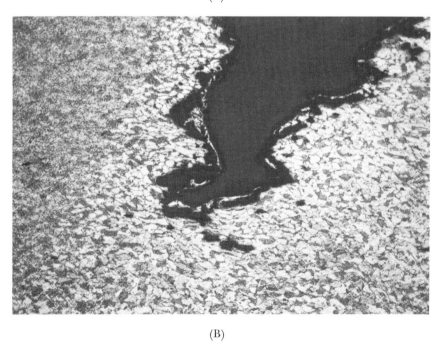

(B)

Fig. 16-19. Creep failure in 1 Cr–½ Mo steel vessel 1⅝-in. thick along metallurgical notch in heat-affected zone where slight difference in metallurgical structure produced a hardness variation of 20 points Brinell. [Magnification (B) 50×]

Since there has been a long history of satisfactory life and performance of 1¼Cr–½Mo boiler tubing and piping at temperatures of 1000 and 1050°F, the change does not appear necessary. Far too often, an isolated failure is utilized as the basis for code or design practice changes. Such changes make particularly little sense when it is proved that the failure was the result of the improper use of the material.

HYDROGEN EMBRITTLEMENT AND DAMAGE

Definition

Hydrogen embrittlement, also referred to as *hydrogen brittleness* represents the loss of ductility of a steel by the absorption of hydrogen. Tensile strength is not significantly reduced. The process is reversible in that ductility can be restored by heat treatment.

Hydrogen damage represents the permanent weakening of the steel caused by the development of microfissures.

The cracking caused by hydrogen damage occurs along grain boundaries. (It differs thus from the hydrogen-induced cold cracking which generally tends to be transgranular.)

Hydrogen embrittlement has occurred in boiler tubing in a substantial number of boilers and in refinery equipment involved in hot hydrogen service such as catalytic reformers. A brittle failure is shown in Fig. 16-20. These failures generally occur at service temperatures between 600 and 1000°F.

A vast amount of information has been published on hydrogen in steel and hydrogen embrittlement in steel base metals and weld deposits. In recent years, considerable information has been published, also, on hydrogen embrittlement and hydrogen damage failures in boiler tubing[25-36] and refinery equipment.[37-41]

Failure Progression

Failures in boiler tubing caused by hydrogen damage generally seem to progress in the following sequence:

(1) localized corrosion (oxidation)
(2) build-up of scale
(3) decarburization in steel underneath the scale surface

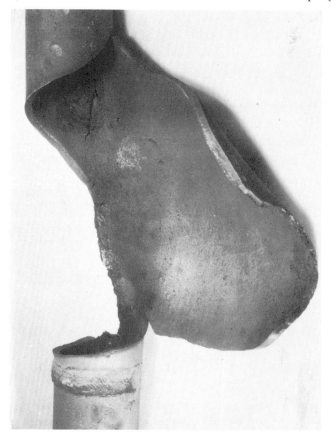

Fig. 16-20. Failure in boiler tube as a result of hydrogen embrittlement and damage.

(4) intergranular microfissuring
(5) major cracking
(6) failure

An example of the scaling and corrosion of the inside tube surface normally accompanying hydrogen embrittlement and damage is shown in Fig. 16-21.

The initial corrosion is generally associated with oxidation. Bonded oxygen (H_2O) [33] and dissolved oxygen [34] have been considered responsible. Under specific operating conditions, bonded or dissolved

Fig. 16-21. Cross section of boiler tube showing localized scaling which preceded hydrogen embrittlement. (Magnification 6×)

oxygen, alone or together, can probably result in the localized oxidation of steel.

The initial oxidation, possibly the result of heat treatment, produces a thin, more or less uniform iron oxide layer along the inside tube surface. As this layer is tightly adherent, it is considered to inhibit further oxidation. With increasing thickness, the rate of oxidation becomes negligible and in effect, stops. Even most of the small corrosion oxidation pits appear to become stabilized without further attack of the steel. It may be that periodic pickling or specific concentrations of acid, salt or caustic loosen the scale in some of the localized pit areas, exposing fresh metal surface to further oxidation. As the pit sharpness increases, the cavity in the metal surface might not be able to accommodate the iron oxide that forms subsequently, since the lower density of the iron oxide requires more space than is available. This tends to result in cracking of the scale layer and in the exposure of more fresh steel surface. Thermal fatigue, mechanical vibrations or acid pickling also may loosen the scale in the larger pit areas. Thus, after a threshold size of cavity has developed along the inside tube wall, the rate of further oxidation and scale accumulation may become substantial.

358 Defects and Failures in Pressure Vessels and Piping

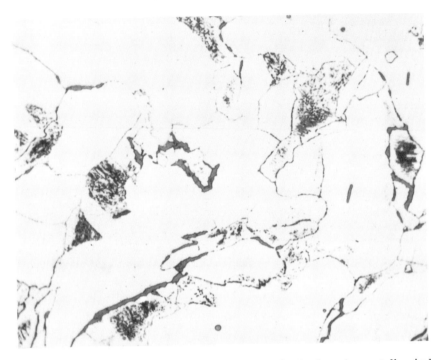

Fig. 16-22. Photomicrograph confirming decarburization in metallurgical structure and cracking along grain boundaries in boiler tube which failed by hydrogen embrittlement and damage. (Magnification 1000×)

Decarburization of the type illustrated in Fig. 16-22 is generally associated with the occurrence of hydrogen damage in steel.

Hydrogen Diffusion Into Steel

It has been recognized for a long time that hydrogen which readily enters molten steel or weld metal may also diffuse into steel at room temperature. Thus, if an atomic hydrogen layer is developed at room temperature on the steel surface, as occurs readily in electrolytic (electroplating) reactions, the hydrogen tends to diffuse into steel.

Hydrogen diffusion data are usually based on permeability characteristics. Although diffusion of atomic hydrogen through metals tends to increase with increasing temperature, permeability may vary with the method of introducing the hydrogen. Nascent hydrogen released at the liquid-to-metal interface, as in an electrolytic cell, may introduce large volumes of hydrogen into the metal. This tends to develop high stresses along the metal interface developing a high permeability. The permeability is also influenced by cold work and other earlier treatments of the steel. Iron previously subjected to hydrogen penetration is more readily permeable on each subsequent exposure. The increased permeability of freshly pickled steel is an example of this effect. Conditioning heat treatments may restore the steel to its original permeability.

The atomic hydrogen diffusing into the steel surface does not necessarily result in damage or even embrittlement although it is believed that in voids, the atomic hydrogen combines into molecular hydrogen. The formation of molecular hydrogen results in severe localized stresses. Where the steel exhibits sufficient ductility, it can withstand these internal localized stress concentrations. However, if the steel is hard and brittle, or if the stress system does not permit plastic deformation, microfissuring with subsequent failure may occur. Thus, the degree of embrittlement is normally considered to depend on the amount and nature of the hydrogen present, the magnitude, distribution and characteristics of the internal stresses, and the ability of the material to withstand these stresses. As the severity of the internal stress system is increased, less hydrogen is required to cause embrittlement. These considerations explain why cracking as a result of hydrogen embrittlement is normally rarely observed in mild steels and in austenitic stainless steels. It is far more common in low-alloy steels of high hardenability.

Methane Formation

Under certain conditions, however, mild steel tends to become severely embrittled by hydrogen leading to permanent hydrogen damage. In boiler tube failures, this is associated with the decarburization of the iron-carbide (Fe_3C) particles in steel by hydrogen and the formation of methane.

Methane is formed by the reaction:

$$Fe_3C + 2H_2 \rightleftarrows CH_4 + 3Fe$$

The reaction itself is reversible, depending on the temperature and pressure involved, although any crack formation cannot be reversed at this stage. At each temperature and pressure, a specific quantity of methane is in equilibrium with a specific quantity of hydrogen. At approximately 875°F, the action is normally considered to be in equilibrium. At lower temperatures, the reaction tends to go to the right and form methane, while at higher temperatures, it tends toward hydrogen formation. Thus, under equilibrium conditions if the temperature is increased, the methane will decompose and produce hydrogen which restores equilibrium conditions. Actually, in steel, the "reversal" temperature seems to be approximately 200°F higher due to the fact that thermodynamic equilibrium conditions are not obtained.

The cracking illustrated in Fig. 16-22 generally occurs initially along the grain boundaries of steel. In a manner similar to the graphitization failures, the iron carbide (Fe_3C) in the pearlite grains decomposes. The carbon atoms then must diffuse toward the grain boundaries. The hydrogen in the grain boundary voids combines with the carbon atoms to form the methane which is unable to diffuse out of the steel. With increasing methane pressure, fissuring along the grain boundaries takes place resulting in nonreversible hydrogen damage.

Effects of Time and Temperature

Whereas the tendency toward methane formation increases with lower temperatures, the rate of carbon diffusion decreases. Thus,

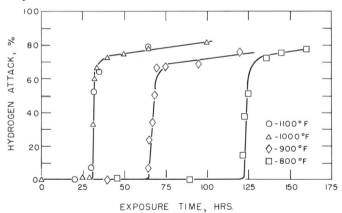

Fig. 16-23. Hydrogen attack of steel as a function of exposure time at 700 psi hydrogen pressure.[42]

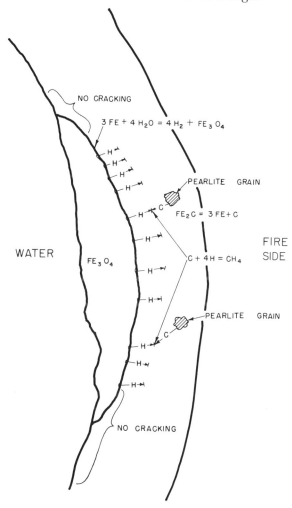

Fig. 16-24. Sketch illustrating mechanism involved in hydrogen embrittlement and damage in boiler tubing.

at a sufficiently low temperature, the reaction involving decarburization does not take place. For example, decarburization has not been observed in mild steel in service at room temperature. In mild steel involved in prolonged high temperature service, iron-carbide decomposition and carbon diffusion toward the grain boundaries normally has not been observed at temperatures below 800 to 825°F. Even at

800 to 850°F, very long periods of time are involved in this diffusion. In the presence of relatively high concentrations of hydrogen in steel, and particularly molecular hydrogen in voids resulting in high internal stresses, the threshold temperatures for carbon diffusion, however, tend to be reduced below 800°F. Moreover, substantially increased rates of carbon diffusion appear to be the result.

Weiner [42] has also shown experimentally that hydrogen embrittlement can occur in relatively short periods, Fig. 16-23, at temperatures between 800 and 1100°F. Once an initial "incubation" period has elapsed, the embrittlement occurs very rapidly. The incubation period was shortest between 1000 and 1100°F, producing embrittlement in as little as 25 to 30 hours. Probably, the higher rates at 1000 to 1100°F as compared to 800°F can be associated with greater mobility of carbon. Weiner showed also that increasing the pressure decreases the incubation time further.

Another factor complicating analysis involves the varying temperatures which occur from boiler start-up to normal operation. These cause variations in the different reactions and diffusion processes. Moreover, the heavy scale on the tube inside the corroded areas tends to complicate the analysis.

On the basis of the various factors discussed above, the mechanism sketched in Fig. 16-24 may be considered as representing the type of failures which occur in boiler tube sections.

Localized corrosion occurs as a result of oxygen being liberated at the steel surface. This may have been caused by acid pickling having entered the boiler. (Salt water pumped through boiler tubing has been found also responsible.) As a result of the corrosion, an Fe_3O_4 scale is formed over the corroded areas. As long as the scale is relatively thin or small, hydrogen formation and embrittlement probably do not become a factor.* However, as the thickness and size of the scale deposit increase, it begins to act as an insulator to the heat transfer across the boiler tube wall. This tends to increase the steel tem-

---

* This differs from the hydrogen embrittlement reported by Kaufman, Trautman, and Schnarrenberger [25] in screen tubing. They contend that the iron oxide scale was washed away continuously, exposing fresh steel surface to the action of water. The bonded oxygen in the water reacting with iron to form more oxide and generating atomic or nascent hydrogen which then penetrates into the steel surface.

Service Failures—Stress and Fatigue 363

perature in the scaled area. At a certain threshold temperature level, the reactions in the steel involving decarburization and methane formation then become significant. Since the temperature differential underneath the scale deposit decreases toward the edge of the scale and corroded area, less embrittlement and finally, none would be expected toward the edge of the corroded areas.

The threshold temperature required to cause significant embrittlement will probably vary, and will be somewhat lower when increased quantities of atomic hydrogen are produced along the metal surface. The entrapment of pickling solutions (or salt) underneath heavily scaled areas then are likely to enhance the hydrogen embrittlement. Because of the decarburization which seems to have occurred in every embrittled specimen to date, it is believed that the temperatures where carbon diffusion becomes significant and methane tends to form are somewhere near 600°F as the lower limit and probably 1000°F as the upper limit. Although carbon steel in contact with hydrogen (as in hydrogenation plants) has failed at temperatures as low as 500°F, only a few isolated instances of failures in boiler tubing have been reported at temperatures below 600°F.

The influence of acid cleaning also does not seem clear. Acid cleaning should be beneficial by removing heavy scale. This, however, will expose fresh steel surface which, under some conditions, becomes susceptible to oxidation. Tube failures due to hydrogen embrittlement have occurred in boiler tubing frequently acid pickled. Conceivably, it may be argued that the pickling did not result in removal of the scale deposit. The acid, by becoming entrapped between the scale and the steel surface may, in fact, increase the formation of nascent hydrogen.

Effects of Heat Treatment and Alloying

By stabilizing the iron carbide (Fe_3C) in steel, the tendency toward methane formation and microcracking can probably be reduced. However, in carbon steels, tempering results primarily in spheroidization of the iron-carbide particles which, although probably beneficial, is not likely to inhibit the decomposition of iron carbide in the presence of hydrogen. This is supported by experimental results by Weiner [42] who showed in test exposures at 900°F that the incubation period for embrittlement was increased from about 55 hours for no heat treatment to 120 hours for a 10-hour heat treatment at 1310°F. A 40-hour heat treatment at 1310°F extended the incubation period to 300 hours. This response, of course, would vary also with each commercial heat of boiler tube steel. Carbide stabilization by alloying with molybdenum would be more effective, provided the steel has been heated for 4 hours at about 1300°F. This heat treatment has effectively inhibited carbide decomposition in environments causing graphitization by forming stable iron-molybdenum carbide. Of course, alloying with chromium, tungsten, and vanadium, or other stronger carbide-forming elements such as titanium is even more effective.[43] Steels alloyed with Cr, Cr-Mo, CrMoV and CrMoVW are now utilized in Germany in boiler tube service when hydrogen embrittlement and damage are suspected.[31]

In the selection of materials for hydrogenation plant service, the Nelson chart shown in Fig. 16-25 is extensively used as a guide.[45] This chart predicts the safe operating limits of temperature and hydrogen pressure of carbon and low alloy steels.

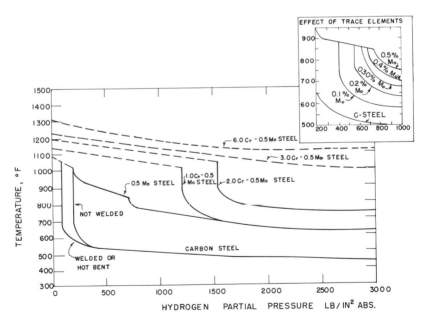

Fig. 16-25. Nelson curve on safe operating limits of temperature and hydrogen pressure for carbon and low alloy steels in refinery service.[45]

HYDROGEN BLISTERING

Hydrogen blistering is caused by the diffusion of atomic hydrogen into the steel. Normally, the hydrogen tends to diffuse through the steel. However, when reaching a void, lamination or slag, or sulfide inclusion, the atomic hydrogen changes to molecular hydrogen. The continued penetration of atomic hydrogen into the void spaces results in the continued formation of more molecular hydrogen. This produces high internal pressures which have reached levels as high as several thousand psi. This may result in actual splitting of the steel. The internal splitting usually appears in the form of blisters on the steel surface. Occasionally, serious rupturing of the steel has occurred.

Hydrogen blisters are particularly prone to occur in "dirty" steels containing significant levels of inclusions, laminations, etc.

Hydrogen blistering has been quite common in catalytic reformers and has been reported at temperatures below 600°F.[46]

Hydrogen blistering has occurred also in butane storage tanks 36 ft in diameter.[47] The butane was run down from the gas separation and treating plant at a temperature of approximately 70°F. Occasionally, the butane was contaminated by small quantities of water with minute quantities of H_2S. Normally, when a liquid water phase is present, corrosion liberates molecular hydrogen which does not enter the steel. However, in the presence of H_2S, molecular hydrogen is not readily formed, resulting in an increased concentration of atomic hydrogen and its diffusion into the steel.[47]

Sandblasting of the surface followed by painting with an amine cured epikote paint has been one method successfully used to eliminate blistering.[47]

OXIDATION INSIDE CRACKS

Oxide scale inside of cracks may also contribute to crack propagation since it tends to resist the normal thermal contraction when the metal cools.

It must also be recognized that some failures in high temperature piping systems may actually involve initiation or propagation of cracking during shutdown periods when the piping is at low tem-

366   *Defects and Failures in Pressure Vessels and Piping*

peratures. Static stresses are likely to become increasingly severe as the piping system cools and contracts.

EXCESSIVE SERVICE CONDITIONS OR INADEQUATE DESIGNS

Lack of design know-how and improper specification requirements contribute to service failures. Many pressure vessels and piping systems are designed by engineers with insufficient awareness of the effects of service conditions on the system. The effects of thermal and mechanical fatigue are often ignored or inadequately considered. Moreover, requirements are often written into specifications that will result in mechanical or metallurgical notches of substantial magnitude. Thus, even though the service environment is often considered the primary cause of a failure, better know-how in the preparation of design and specifications might have avoided the failure.

Fig. 16-26. Set-up for thermal fatigue testing to evaluate effects of design and weld defects on crack initiation and propagation.

With increasingly severe environments, service failures will continue to have more disastrous consequences. Far more testing is required to evaluate the resistance to failure by thermal or mechanical fatigue.

The thermal fatigue test set-up in Fig. 16-26 is used to evaluate how critical various joint factors and weld and base metal defects are. In the particular test shown, the pipe section has a butt weld with various defects in the root weld pass. It is subjected to heating to 1000°F and water quenching to below 200°F to produce thermal fatigue. Some defects have not propagated and produced failure after 5000 to 10,000 cycles. Others have propagated and caused failure in as few as 100 cycles.

REFERENCES

1. Yao, J. T. P. and Munse, W. H., "Low-Cycle Fatigue of Metals—Literature Review," *Welding J.*, 41, Res. Suppl., 182s–192s (1962).
2. Sachs, G., Gerberich, W. W., Weiss, V., and Latorre, J. V., "Low-Cycle Fatigue of Pressure-Vessel Materials," *Proc. ASTM*, 60, 512–529 (1960).
3. Kooistra, L. F. and Lemcoe, M. M., "Low-Cycle Fatigue Research on Full-Size Pressure Vessels," *Welding J.*, 41, Res. Suppl., 297s–306s (1962).
4. Sessler, J. G. and Weiss, V., "Low-Cycle Fatigue Damage in Pressure-Vessel Materials," *Trans. ASME*, 85, Series D, 539–547 (1963).
5. McClure, G. M., Eiber, R. J., Hahn, G. T., Boulger, F. W., and Masubuchi, K., "Research on the Properties of Line Pipe," Summary Report Catalog No. 40/PR, 1962, American Gas Association, New York.
6. International Institute of Welding "Reports on Fatigue Failures—Fracture of a Carbon Monoxide Regenerator in Stainless Steel," Report IIS/IIW-90-62 of Commission XIII, Welding Research Abroad, 9, 86–87 (Dec. 1963).
7. Bonin, C. C., "Water-Hammer Damage to Oigawa Power Station," *Trans. ASME*, 82, Series A, 111–116 (1960).
8. Harris, H., "Metallurgical Investigation of Failure of a Boiler During Hydrostatic Test at Sizewell Nuclear Power Station," Special Report of West of Scotland Iron and Steel Institute, 1964.
9. Thielsch, H., "Thermal Fatigue and Thermal Shock," Welding Research Council Bulletin No. 10 (Apr. 1952).
10. Edison Electric Institute "1950 Report of the Metallurgy and Piping Subcommittee of the Prime Movers Committee."
11. Gatewood, A. R., "A Steam Line Failure and an Investigation of Shafting Material," Report to Gulf Section Soc. Nav. Architects and Marine Engrs., American Bureau of Shipping (Apr. 20, 1951).
12. Edison Electric Institute Memoranda to Prime Movers Committee "Hot Reheat Line Failure"—Ashtabula No. 5 (1961).
13. Coulter, S. B. and Jackson, R. L., "Operating of Steam Turbines to Minimize Shell Cracking," *Trans. ASME*, 82, Series A, 227–245 (1960).

14. Jackson, R. L., Coulter, S. B., and Sheppard, R., "Importance of Matching Steam Temperatures With Metal Temperatures During Starting of Large Steam Turbines," *Trans. ASME,* **79,** 1669–1678 (1957).
15. Wyatt, L. M. and Gemmill, M. G., "Experience With Power Generating Steam Plant and Its Bearing on Future Developments," Proceeding of the Joint International Conference on Creep, London, 1963.
16. Tietz, H., Buchholtz, H., Werner, M., Ruttmann, W., and Schinn, R. "Ein Höchsttemperatur-Kraftwerk mit einer Frischdampftemperatur von 610°C," *Zeit. Verein. Deutsch. Ing.,* **95,** 801–831 (1953).
17. Ruttmann, W. and Brunzel, N., "10 Jahre austenitischer Stähle im Kesselbetrieb," *Mitt. Ver. Grosskesselbesitzer,* **80,** 310–326 (1962).
18. Verein. d. Grosskesselbesitzer, 1958/59 Review.
19. Private Communication, Bayernwerk, A. G., Germany.
20. Smith, W. R., "Report of the Cause of Failure and Repair of the VBWR Recirculation Piping," Atomic Power Equipment Department, General Electric Co., 1962.
21. Verein. d. Grosskesselbesitzer, 1957/58 Review.
22. Verein. d. Grosskesselbesitzer, 1956/57 Review.
23. Rutherford, J. J. B., "Some Experiences in Service—Power, Oil and Chemical Plants," *High Temperature Properties of Metals,* The American Society for Metals, Cleveland, 1951.
24. Buchanan, J. R., "Accidents in Nuclear Energy Operations," *Nucl. Safety,* **3,** 90–96 (June 1962).
25. Kaufman, C. E., Trautman, W. H., and Schnarrenberger, W. R., "Action of Boiler Water on Steel—Attack by Bonded Oxygen," *Trans. ASME,* **77,** 423–430 (1955).
26. Corey, R. C., "Corrosion of High-Pressure Steam Generators; Status of our Knowledge of the Effect of Copper and Iron-Oxide Deposits in Steam Generating Tubes," *Proc. ASTM,* **48,** 907–942 (1948).
27. Grabowski, H. A., "Corrosion of Steel in Boilers—Attack by Dissolved Oxygen," *Trans. ASME,* **77,** 433–441 (1955).
28. Hankison, L. E. and Baker, M. D., "Corrosion and Embrittlement of Boiler Metal at 1350 Pounds Operating Pressure," *Trans. ASME,* **69,** 479–486 (1947).
29. Mitsch, E. H. and Yeager, B. J., "Experience With Internal-Boiler-Surface Corrosion in 1450-Lb Open-Pass Boilers at West End Station of the Cincinnati Gas and Electric Company," *Trans. ASME,* **69,** 487–491 (1947).
30. Straub, F. G., "Wall-Tube Corrosion in Steam-Generating Equipment Operating Around 1300 psi," *Trans. ASME,* **69,** 493–499 (1947).
31. Class, I., "Stand der Kenntnisse über die Eigenschaften Druckwasserstoffbeständiger Stähle," *Stahl u. Eisen,* **80,** 1117–1135 (1960).
32. Powell, S. T. and Von Lossberg, L. G., "Unusual Corrosion in High Pressure Boilers," *Corrosion,* **5,** 71–78 (1949).
33. Dick, I. B., "Experiences With Hydrogen Embrittlement in the Consolidated Edison System," *Trans. ASME,* **86,** Series A, 327–340 (1964).
34. Galloway, E. E., "Hydrogen Damage in a 1250-Psi Boiler," *Trans. ASME,* **86,** Series A, 341–343 (1964).
35. Esposito, O. J. and Harrington, R. T., "Case History: Hydrogen-Damage Experience in a Boiler," *Trans. ASME,* **86,** Series A, 299–304 (1964).

36. Partridge, E. P., "Hydrogen Damage in Power Boilers," *Trans. ASME,* **86,** Series A, 311–324 (1964).
37. Nelson, G. A. and Effinger, R. T., "Blistering and Embrittlement of Pressure Vessels by Hydrogen," *Welding J.,* **34,** Res. Suppl., 12s–21s (1955).
38. Manuel, R. W., "Hydrogen Service Failures of Vessels With Insufficient Alloy Content," *Corrosion,* **17,** 435t–436t (1961).
39. Ciuffreda, A. R. and Rowland, W. O., "Hydrogen Attack of Steel in Reformer Service," *Proc. API Div. of Refining,* **37** (III), pp. 116–128 (1957).
40. Vitovic, F. H., et al., "Rate of Irreversible Hydrogen Attack of Steel at Elevated Temperatures," *Proc. API Div. of Refining* (1961).
41. Cooper, C. M., "Material Deficiencies in Welds of Hot Hydrogen Process Equipment Cause Serious Failure," *Materials Protection,* **3,** 34–40 (Jan. 1964).
42. Weiner, L. C., "Kinetics and Mechanism of Hydrogen Attack of Steel," *Corrosion,* **17,** 137t–143t (1961).
43. Neumann, F. K., *Stahl u. Eisen,* **57,** 889–899 (1937).
44. Goodger, A. H., "Corrosion Fatigue Cracking Resulting from Wetting of Heated Metal Surfaces, with Special Reference to Steam Power Plant," *Proc. Inst. Mech. Eng.,* **171,** 394–400 (1957).
45. Nelson, G. A., "Hydrogenation Plant Steels," *Proc. API Div. of Refining,* 29M (III) pp. 163–174 (1949).
46. Resume of Investigations on Steel for High-Temperature High-Pressure Applications 1955–1956, The Timken Roller Bearing Co., Canton, Ohio.
47. Pull, D. J., "Mechanical Engineering in a Modern Oil Refinery," *Proc. Inst. Mech. Eng.,* **176,** 495–521 (1962).

chapter 17

SERVICE FAILURES—CORROSION

TYPES OF CORROSION

Corrosive failures due to excessively severe service conditions which have not been anticipated will be of primary concern in this chapter. By definition, corrosion is the destruction of a metal by chemical or electrochemical reactions with its environment.

Many types of corrosive mechanisms are recognized that can cause or contribute to pressure vessel, tank, and piping failures. A number of examples of failures in which corrosion was involved have been discussed in previous chapters in which corrosion was considered as a contributing factor rather than as the primary cause.

Corrosion failures abound when conditions become more severe than was anticipated at the time the pressure vessel, tank or piping system was designed or when the materials were originally selected. A service environment may become more corrosive as a result of changes in the chemical environment or in the solutions contained in the vessel or tank, or carried in the piping. Conditions may also arise that substantially reduce the corrosion resistance of the materials. An example of this would be the formation of heavy copper deposits on the inside of steel tanks or pipe, resulting in galvanic corrosion.

While such changes should be anticipated, and accounted for by changes in design, materials or operation, the fact is that they frequently are not.

Service Failures—Corrosion

This discussion is not primarily concerned with corrosion as such, which is adequately covered in various texts and handbooks.[1,2] Of interest here are those corrosive failures resulting from excessive operating conditions. Furthermore, it is not practical to document all the different types of corrosion failures encountered. The examples presented here illustrate those that can be expected in pressure vessels, tanks, and piping components.

Many failures caused by corrosion are not easy to classify. Various interrelated mechanisms can be involved. Changes in service conditions can further complicate failure analysis.

The more common types of corrosion failures associated primarily with service conditions are:

(1) general corrosion
(2) galvanic corrosion
(3) corrosion pitting
(4) intergranular corrosion
(5) knife-line corrosion
(6) crevice corrosion
(7) stress-corrosion cracking and corrosion fatigue
(8) stress-enhanced corrosion
(9) erosion
(10) exfoliation
(11) corrosion by soil or insulation
(12) scaling and oxidation

The consequence of the corrosive deterioration normally represents four different types:

(1) *General corrosion* where an over-all deterioration occurs. In this type of attack, surface changes usually tend to be gradual or uniform. Such an attack may or may not involve the formation of corrosion products.

(2) *Localized corrosion* with no attack or only minor corrosion occurring in the same general over-all area of the surface. This type of attack may involve pitting or crevice corrosion.

(3) *Mechanical-chemical corrosion* involving the action of stresses and chemical effects. This type of attack often occurs in combination with cracking or severe localized corrosion.

(4) *Erosion* attack involving primarily mechanical wear by the

impact of liquids, slurries, or steam usually at a high velocity. Chemical attack may be a contributing factor, for example, by removing protective films on the metal surface.

GENERAL CORROSION

General corrosion is characterized by a relatively uniform attack over the vessel surface or around the pipe circumference. This may occur throughout a vessel, tank, or piping system or in a specific section, such as the bottom of a tank.

Figure 17-1 illustrates general thinning in a vertical carbon steel pipe section near the end of the system, just before entering a flash tank. The hot condensate, when not flowing, contained dissolved carbon dioxide, or carbonic acid, which caused the relatively uniform attack. The corrosion was more severe in the rimmed steel pipe (ASTM A53) below the weld than in the silicon deoxidized steel pipe (ASTM A106) above the weld. Deaeration of condensate in an isolated area has also been responsible for this type of differentiating corrosive behavior.

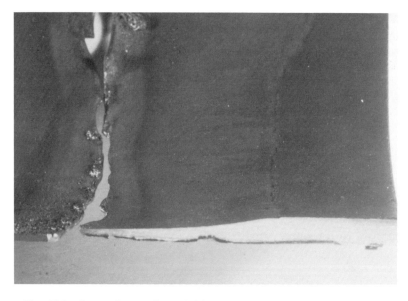

Fig. 17-1. General corrosion of 6-in. OD carbon steel pipe in hot condensate return system caused by excessive carbon dioxide in condensate.

GALVANIC CORROSION

Galvanic corrosion is caused by the current of a galvanic cell made up of dissimilar electrodes, such as iron and copper, or brass and aluminum. Such corrosion may occur where dissimilar metals are joined by welding or brazing, riveting, or mechanical forming. It can also occur on surfaces involving plated, sprayed, or other metal coatings or some paints applied either inadvertently or by design.

An example of inadvertent metal coating involves the plating out of a heavy continuous layer of copper on the inside of steel pipe. Figure 17-2 illustrates corrosion in boiler feed pump discharge piping. A cross section of the edge of the corroded cavity confirmed that a copper layer had been deposited on the inside of the pipe surface and that the steel was corroding as a result of galvanic action, as shown in Fig. 17-3. The boiler feed water, in passing through heat exchangers and other equipment containing tubing of copper or

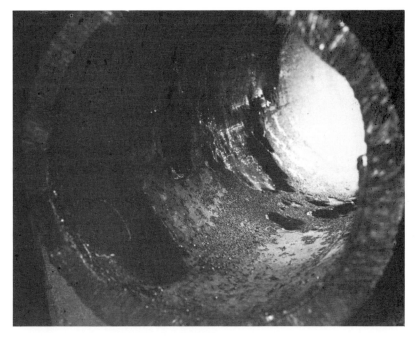

Fig. 17-2. Galvanic corrosion in 6-in. nominal OD Schedule 120 carbon steel boiler feed pump discharge piping system.

374 *Defects and Failures in Pressure Vessels and Piping*

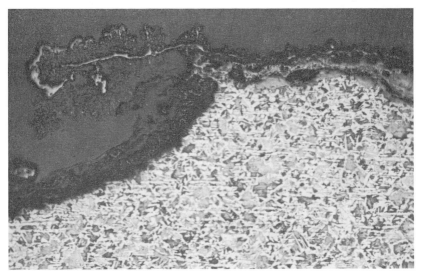

Fig. 17-3. Copper layer on steel surface at edge of corroded cavity on inside of steel pipe. (Magnification 100×)

high copper alloys, very likely picked up soluble copper and copper oxides. Dissolved oxygen, carbon dioxide, and possibly sulfur dioxide, are likely sources providing the oxidation potential to convert copper to a soluble form.[3] Under slightly acid pH conditions, copper tends to plate out on the steel surface.

During acid cleaning, iron oxides on steel surfaces present as FeO, Fe_2O_3, and Fe_3O_4 are removed (dissolved), forming the iron chlorides $FeCl_2$ and $FeCl_3$. When soluble copper is present, the copper tends to displace the iron in the $FeCl_3$, forming the soluble copper salt. This solution tends toward an equilibrium in favor of the reduction of ferric iron to ferrous iron, with the resultant formation of metallic copper.

If the copper salts are not removed prior to the removal of the iron oxides with acids, steel surfaces may be flash plated with a layer of copper during the acid state. Although proper treatment of the boiler water may be employed subsequently, preventing further plating out of copper, the resulting layer of copper, if sufficiently large and continuous, can do harm and promote galvanic corrosion.

It is possible, of course, that the copper flash plating occurred during one of the boiler cleaning operations when the acids and re-

Fig. 17-4. Galvanic corrosion in 6061 aluminum condenser tubing along the edge of a paint film and around paint droplets.

actions were not in balance, resulting in the formation of excessive copper salts susceptible to plating out.

The subject of copper deposition in pipe and boiler tubing is quite controversial. In numerous instances, copper deposition is not harmful. In many power plants, copper and CuO are virtually collected by the barrels in mud drums and atemporators without any evidence of galvanic corrosion in the respective piping and tubing systems. In boiler tubing, copper is frequently observed deposited on iron oxide, usually in the form of small, integral particles. This has not caused trouble in high pH environments, except to reduce heat transfer slightly.

An example of a paint containing iron oxide causing galvanic corrosion in 6061 aluminum condenser tubing is illustrated in Fig. 17-4. The paint was sprayed inadvertently into the $3/4$-in. diameter by 0.047-in. wall tube ends. Galvanic corrosion occurred along the edge of the paint film and around droplets of paint adhering to the tube surface. Lake water was the "electrolyte" flowing through the tubing.

Minor Dissimilarities in Composition

Even with minor dissimilarities in composition, the presence of different metallurgical phases can lead to galvanic corrosion in some service environments. One of the structural components will be corroded in favor of another. An example of the corrosive attack of the ferrite phase in an austenitic matrix was shown in Fig. 4-17.

Attack of the ferrite phase in a Type 308L stainless steel weld deposit in monovinyl chloride service is illustrated in Fig. 17-5.

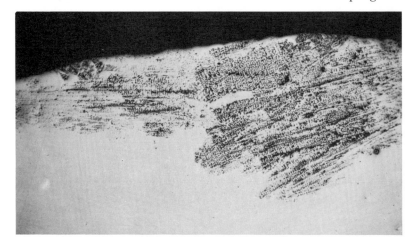

(A)

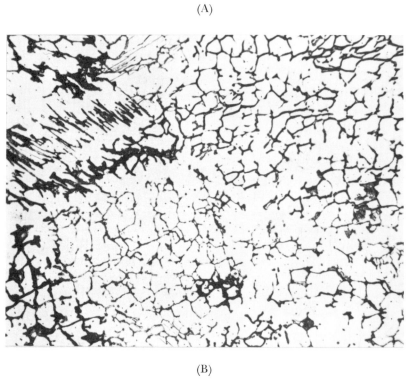

(B)

Fig. 17-5. Attack of ferrite phase in Type 308L stainless steel weld deposit in 2-in. by 0.154-in. Type 304L stainless steel pipe carrying monovinyl chloride. [Magnifications (A) 50× and (B) 500×]

INTERGRANULAR CORROSION

Intergranular corrosion is generally associated with stainless steel materials, such as Types 304, 309, 310, 316 or 317 that have been heated to, or held at, temperatures between 900 and 1500°F and have been subsequently exposed to corrosive environments that attack these grades along the grain boundaries. The attack can occur quite rapidly. For example, Type 317 stainless steel heat exchanger tubes containing 3 to 4% molybdenum have failed after only 2 weeks of service in a sulfur dioxide water separator in a gasoline-treating plant.[4]

The susceptibility to intergranular corrosion is caused by the formation of a complex chromium carbide that depletes the chromium content along the grain boundaries. Because of this lowering of the chromium content along the grain boundaries, the steel is also called *sensitized*. Examples of this type of attack have been shown in previous chapters with various primary causes of failure: improper selection of materials (Fig. 4-14), lack of required heat treatment (Fig. 14-4), and incorrect pickling (Fig. 15-2). Changes in service conditions can be responsible for this type of corrosive attack when the acid concentration is changed to a level at which stainless steel materials are attacked intergranularly.

If the service requirements necessitate such a change, the best remedy is to replace the vessel or piping with a material not susceptible to intergranular corrosion in the specific service environment. Stabilized stainless steel Types 321 or 347 or the extra-low-carbon grades Types 304L or 316L with 0.03% maximum carbon content may be suitable. Specific heat treatments may also improve corrosion resistance. In many instances, however, heat treatments are not practical for pressure vessels, tanks or piping components because of the distortion resulting from uneven heating and cooling.

KNIFE-LINE CORROSION

Knife-line corrosion occurs along the plane between base metal and weld metal. An example in a Type 316 stainless steel pipe weld is illustrated in Fig. 17-6.

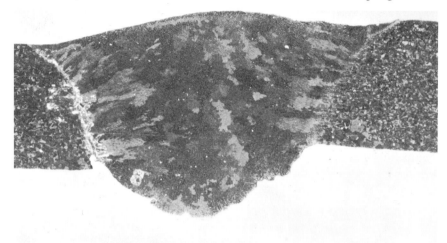

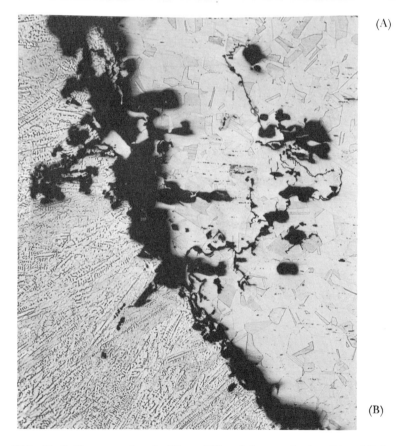

Fig. 17-6. Knife-line corrosion in Type 316 stainless steel pipe 8-in. OD by 0.109-in. wall in hot water service containing approximately 60 ppm chlorides. [Magnifications (A) 13×, (B) 100×]

Knife-line corrosion is most commonly observed in austenitic stainless steels, particularly those grades containing molybdenum. It has occurred also in high nickel alloys containing molybdenum.

Knife-line attack is generally associated with the formation of complex carbides at temperatures near the melting point or with the presence of low-melting phases in the base metal adjacent to the weld which promote localized galvanic corrosion.

PITTING CORROSION

Corrosion involving localized pitting represents a type of attack extensively encountered in ferrous and nonferrous materials.

Pitting, as most generally encountered, is also described as *concentration-cell* corrosion, as it results from the electrochemical potential set up by differences in oxygen concentration within and outside the pit. The oxygen-starved pit acts as anode and the unattacked metal surface as cathode.

Pitting corrosion is most likely to occur during operational outages, when stagnant no-flow conditions exist. These provide the environments most suitable for build-up of concentration cells.

The susceptibility to pitting corrosion is further enhanced by scratches, surface defects, breaks in protective scale layers or metal surface films, or grain boundary conditions.

Carbon Steel

Pitting has occurred extensively in carbon steel tubing used to heat untreated water containing oxygen.[5]

Pitting as a result of service conditions is also frequently encountered in steam condensate and district heating piping, where excessive oxygen is present in the condensate. An example of pitting is shown in Fig. 17-7.

Chemical treatment of the condensate or design changes to eliminate the solution of oxygen in the condensate should avoid this type of corrosion.

The presence of excessive oxygen in boiler feed water in feed water preheaters during shutdown periods has also caused pitting corrosion, particularly in cold worked and weld joint areas.[6,7] This attack is believed to occur primarily at pH values between 8 and 10.[6]

380 *Defects and Failures in Pressure Vessels and Piping*

Fig. 17-7. Pitting in 2-in. Schedule 40 carbon steel pipe carrying steam condensate.

During operations involving flow conditions, such attack has not been observed. Chemical treatment may also prevent this attack.

Another instance of pitting corrosion in water wall boiler tubing, illustrated in Fig. 17-8, has led to the crack-like failure on the outside of the tube shown in Fig. 17-9. The carbon steel tube 3-in. OD by 0.57-in. wall failed after 22 years at 575°F. The causes of pitting in boiler tubing are not always clearly established due to changes in the operating conditions.

Deposits of metallic copper are quite frequently observed in the vicinity of and within pits and craters. They are generally believed to be the result of the reduction of soluble copper or copper oxide by hydrogen. When, in pitting corrosion, hydrogen is released within the pit, any soluble copper present in the feed water may be reduced to metallic copper. The resulting copper particles will then remain within the pit as so-called "spectator" deposits,[3] Fig. 17-10. It is generally believed that these minute copper particles, produced as a

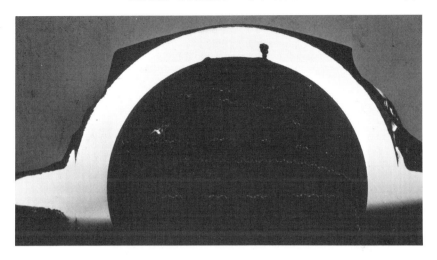

Fig. 17-8. Pitting in 3-in. OD by 0.57-in. thick water wall tube section.

Fig. 17-9. Failure by pitting after 22 years of service at 575°F of 3-in. OD by 0.57-in. thick water wall tube.

382 *Defects and Failures in Pressure Vessels and Piping*

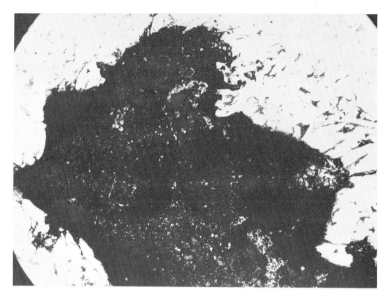

Fig. 17-10. Example of copper particles formed within pit progressing through boiler tube wall. (Magnification 90×)

result of secondary reactions, are of no or only minor consequence in starting or continuing the corrosive attack.[3,8]

Aluminum

Pitting and copper deposition in an aluminum diesel cylinder head is shown in Fig. 17-11. The cause of this attack was the use of an incorrect inhibitor in a copper-nickel-iron tubing system.[9]

Crater-Type Corrosion

Another kind of pitting attack in boiler tubing is known as crater-type corrosion.[3] It occurs on the waterside and is associated with hydroxide alkalinity, Fig. 17-12. This type of attack generally takes the form of a single, bowl-shaped crater, 3/4 in. to 1 1/2 in. in diameter, with steeply sloping sides and a small penetration at the bottom of the bowl. The craters generally tend to be located downstream of a flow disturbance such as a backing ring, poor fit-up or excessive penetration.

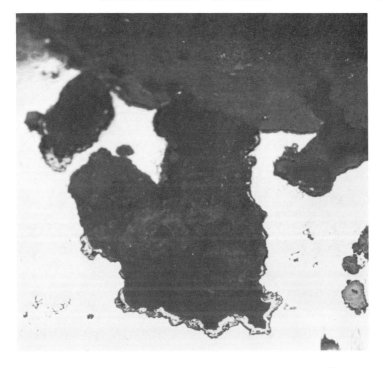

Fig. 17-11. Pitting and copper deposition in aluminum diesel engine cylinder head.[9]

Stainless Steels

Pitting in stainless steels is frequently caused by solutions containing chlorides which tend to penetrate protective surface films. This holds true even for surfaces that have been treated chemically or passivated to improve corrosion resistance. As a result of film breakdown, a pit may form and continue to grow. Figures 17-13 and 17-14 illustrate pitting in Type 304L piping carrying hot water that contained a relatively high chloride ion concentration. If the protective surface film is scratched or worn, localized corrosion may start at the pit nucleus. In some instances, the passivated surface may even act as a cathode, with the pit cavity serving as anode. This results in accelerated corrosion beneath the surface, as shown in Fig. 17-14.

384   Defects and Failures in Pressure Vessels and Piping

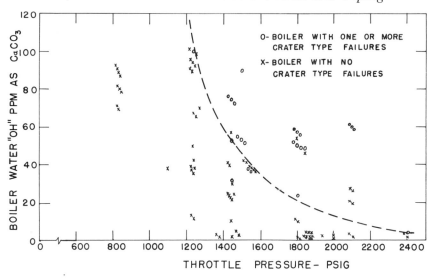

Fig. 17-12. Graph relating pressure versus boiler water hydroxide alkalinity and susceptibility to crater-type corrosion.[3]

Fig. 17-13. Pitting corrosion in Type 304L stainless steel. (Magnification 15×)

Fig. 17-14. Pitting corrosion in Type 304L stainless steel originating at small "pit nucleus." (Magnification 15×)

CREVICE CORROSION

Crevice corrosion is frequently referred to as contact corrosion. It often occurs at the location of contact between nonmetallic materials and passive metals. The corrodent in the crevice is usually depleted of some film-forming constituent as, for example, oxygen. When the protective film on the metal surface is ruptured, corrosion proceeds. Crevice corrosion also generally represents concentration-cell corrosion.

Figure 17-15 shows an example of crevice corrosion. The corrosive attack occurred in a flange of a piping system carrying hot plutonium nitrate-nitric acid solutions. The nitric acid content was of the order of eight molar. Corrosion occurred along the crevice formed by the gasket between the two mating flanges.

Crevice corrosion is also quite common in threaded joints, as shown in Fig. 17-16. This may be prevented by the use of joint compounds or brazing or soldering to prevent the seepage of liquids into the crevice.

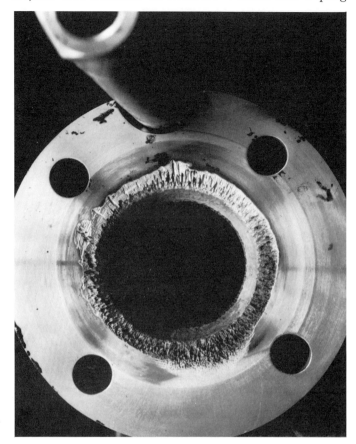

Fig. 17-15. Crevice corrosion on edge of stainless steel pipe in crevice formed by gasket between two mating flanges.

Fig. 17-16. Crevice corrosion in threaded mild steel pipe joint.

Another example of crevice corrosion due to the formation of welding crevices was shown in Fig. 13-9.

STRESS-CORROSION CRACKING

Stress-corrosion cracking is the result of the combined effects of corrosion and tensile stress. Where this occurs, the corrosion alone would not have produced failure by cracking. Conversely, in the absence of corrosive attack, the stresses present in the material would not have produced failure.

Specifically, stress-corrosion cracking is the result of corrosion and static tensile stresses. When cyclic stresses are involved, the process is generally more aptly described as corrosion fatigue.

Stress-corrosion cracking has been widely discussed in the literature [10-12] and has been illustrated elsewhere in this book, (see Index).

For almost every alloy, there appear to be particular environments that render it susceptible to stress-corrosion cracking. For example, carbon steel will undergo stress-corrosion cracking in hot nitrates,[47] hot carbonate waters, and sulfo-nitric acids; austenitic stainless steels in hot chlorides; and copper alloys in ammonia. An example of the last is shown in Fig. 17-17.

The presence of tensile stresses is one of the two principal factors causing stress-corrosion cracking. The stresses may be internal or may be caused by externally applied loads, or both.

Internal stresses can be caused by welding, by uneven cooling from elevated temperatures, or by internal metallurgical transformations involving volume changes. An example of very severe stress-corrosion cracking associated primarily with residual welding stresses in the heat-affected zone of a 2-in. OD by 0.154-in. wall pipe butt weld is shown in Fig. 17-18. Cracking did not occur outside of the heat-affected zone in the Type 304L stainless steel material.

Hot or cold fabricating operations such as forming, bending, upsetting, bolting, riveting, and press or shrink fitting may also introduce residual stresses of significant magnitude. Surface notches can be extremely critical since very severe stresses may be concentrated at the apex. In addition, the notch frequently permits deposition of impurities that may cause corrosion pitting.

In most cases, the internal stresses are of greater practical impor-

388 *Defects and Failures in Pressure Vessels and Piping*

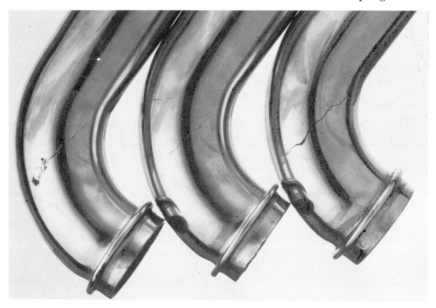

Fig. 17-17. Stress-corrosion cracking in brass tail pieces through which ammonia had been flushed.

tance than external stresses. Since the causes of excessive internal stresses and notches generally relate to improper forming, fabrication, or heat treatments, they have been treated in previous chapters. With few exceptions, operating pressure stresses are sufficiently low to be of comparatively little importance, except as they add to existing internal stresses. Service conditions can be responsible for failure, however, either when the operator introduces internal stresses as a result of handling, bending, hammering, etc., or when the environment is changed to make it more corrosive. Improper chemical cleaning during service may also be responsible for stress corrosion cracking.

Nature of Cracking

Stress-corrosion cracking, like other stress enhanced failures, will start at the location where the tensile stress is the highest and the material's metallurgical structure is the weakest. Thus, localized cold-worked areas, as produced by hammer marks, and localized hard metallurgical structures are particularly prone to start cracking.

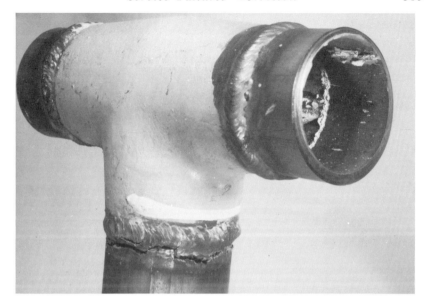

(A)

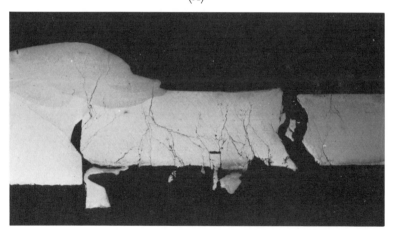

(B)

Fig. 17-18. Severe stress-corrosion cracking of butt welds in Type 304L stainless steel piping. [Magnification (B) 5×]

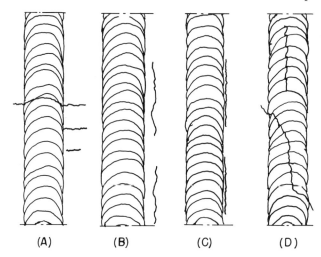

Fig. 17-19. Example of crack patterns due to stress corrosion and corrosion fatigue in butt welds.

Along welds, the cracking may occur in the patterns illustrated in Fig. 17-19. The cracking transverse to the weld shown at (A) is primarily the result of longitudinal weld shrinkage. The cracking illustrated at (B) is caused by transverse shrinkage. The cracking immediately adjacent to the weld bead is generally caused by undercut or hard metallurgical structures. Longitudinal cracking in the center of the weld bead shown at (D) is usually caused by lack of penetration in the weld root.

The loads on branch connection welds and attachment welds may further increase localized stresses and thus affect the location and direction of crack progression. Typical examples are illustrated in Fig. 17-20. Since the corrosive medium is normally on the inside of the vessel or pipe, the cracking normally tends to start from the inside, as illustrated at (G).

Figure 17-21 illustrates a stress-corrosion failure in $3/8$-in. diameter Type 347 stainless steel tubing used as a steam sampling line. Intermittent acid cleaning followed by insufficient rinsing was considered the primary cause of failure. The transgranular nature of this cracking is evident in Fig. 17-22.

Even though equipment may normally operate under conditions and at temperatures where stress-corrosion cracking will not occur,

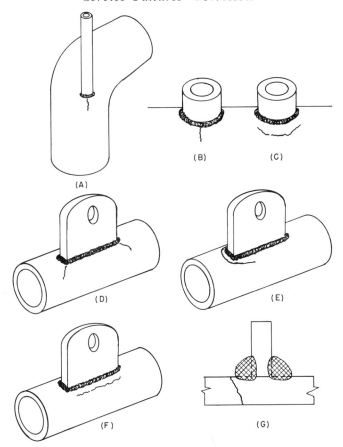

Fig. 17-20. Examples of crack patterns due to stress-corrosion cracking and corrosion fatigue in branch connections and vessel or pipe support attachment welds.

such cracking and failures can be caused during shutdown periods. Figure 17-23 illustrates stress-corrosion cracking on the inside of a 13½-in. OD Type 321 stainless steel pipe. Normal operation involved passage of hot gas at 825 to 930°F and 450 psi from a scrubber to CO converters and heat exchangers. After 3000 hours of operation, failures were found in two 8-in. and 12-in. tee connections and at a branch resulting in gas leakage. The stress-corrosion cracking was caused by unexpected condensate carry-over containing up to 5 ppm chloride during shutdown periods. Carry-over of boiler water

392 *Defects and Failures in Pressure Vessels and Piping*

Fig. 17-21. Stress-corrosion cracking in ⅜-in. diameter Type 347 stainless steel tube section. (Magnification 3×)

containing chlorides into a 10-in. diameter Type 321 superheater steam header carrying steam at 1200°F and 340 psi caused stress corrosion cracking and leaks in 1-in. diameter branch pipe connections into the header, Fig. 17-24. Although thermal fatigue was probably a contributing factor, stress corrosion was considered the primary cause because of the presence of chlorides in the boiler water. Under similar operating conditions, a catastrophic failure occurred in a 12-in. Type 347 stainless steel cast valve body, Fig. 17-25.

In vessels constructed of stainless clad plates, stress-corrosion

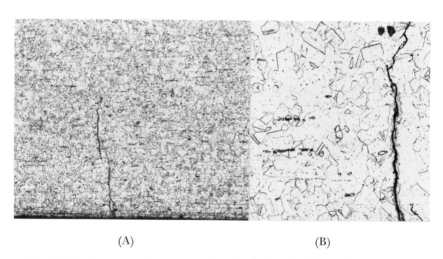

(A)   (B)

Fig. 17-22. Transgranular progression of cracking in Type 347 stainless steel tube. [Magnifications (A) 60× and (B) 300×]

Fig. 17-23. Stress corrosion cracking in 13½-in. OD by 0.7-in. wall Type 321 stainless steel pipe caused by condensate carry-over during shut down periods of hot gas piping system.

cracking will normally stop at the bond between the cladding and backing steel, Fig. 17-26. This is due to the presence of compressive stresses along the bond zone between the stainless steel cladding and the backing steel.

Although most failures due to stress-corrosion cracking start from the inside of vessels and piping, they occasionally start from the outside surface. Fo rexample, at a nuclear reactor installation, stainless steel heat-exchanger tube failures were caused by stress-corrosion cracking induced by chlorides on the exterior surfaces of the tubes exposed to the atmosphere.[13] In another heat-exchanger in a nuclear submarine, Type 347 stainless steel tubing experienced stress-corrosion cracking starting from the secondary side. After 4½ years, the crack penetrated as much as 80% of the wall. The secondary water entered at 110°F and left at 164°F. It normally contained about 5 ppm chloride, but occasionally 15 ppm. The oxygen content was 4 to 6 ppm.[14]

External stress-corrosion cracking of Type 304 and 304L stainless

Fig. 17-24. Stress-corrosion cracking in bore of 1-in. diameter Type 321 branch pipe connection adjacent to fillet weld to 10-in. diameter steam header.

steel heat-exchanger tubing in a nuclear reactor caused by deposits containing substantial chloride ion concentrations was believed caused by either decomposition of chloride-containing gasketing material, or leakage and subsequent evaporation of river water (1 to 6 ppm chloride) through the second tube sheet.[14]

Leaching of chloride from insulation has also initiated stress-corrosion cracking of nuclear piping.[15]

Caustic Embrittlement

A type of stress corrosion is known as *caustic embrittlement*. Specifically, it involves the cracking of a metal under stress and in contact with an alkaline solution. It has been recognized for many decades as a major cause of failure in boilers, particularly those of riveted construction.[16, 17] Some failures have been of the catastrophic type causing severe damage to equipment and buildings.[45]

An example of intergranular progression caused by caustic crack-

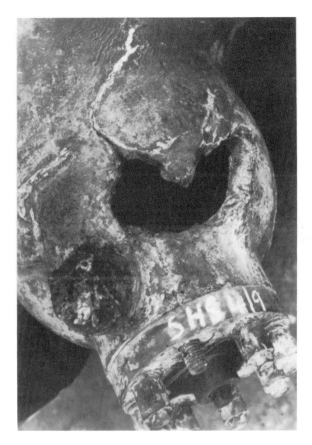

Fig. 17-25. Catastrophic failure of a 10-in. diameter Type 347 stainless steel cast valve body primarily caused by stress corrosion following carry-over of boiler-water.

ing is illustrated in Fig. 17-27. A 40% caustic solution was inadvertently pumped through a carbon steel piping system. Cracking occurred parallel to the weld in the heat-affected zone.

In addition to boiler failures where caustic embrittlement was first recognized, failures have been reported in many other applications. These include gas distribution piping [18] and a 3-ft diameter by 8-ft long preheater drying drum.[19] In the latter failure, one head on the drum blew off, breaking the bearing pedestals and blowing out the adjacent brick wall of the building. Three men were killed.

Fig. 17-26. Cross section through stainless clad steel vessel subjected to stress-corrosion cracking illustrating that cracking did not penetrate into backing steel. (Magnification 50×)

Other examples of caustic embrittlement are illustrated in Fig. 13-22 and in Fig. 14-12.

Caustic is often used in pressure vessels and other equipment utilizing piping to neutralize surfaces previously exposed to acids. In some refinery processes, caustic is also used as an extraction solvent.[20] It may also be used in the manufacture of some greases.

In the heat exchanger of a nuclear reactor, failures occurred in

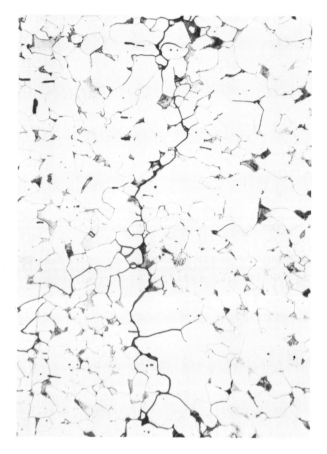

Fig. 17-27. Intergranular stress corrosion resulting from 40% caustic solution inadvertently pumped through a piping system. (Magnification 500×)

Type 304 stainless steel tubing due to free caustic in the secondary water. Although the free caustic content of the water was too low to produce such cracking, it was believed that steam blanketing of an inadequate number of risers allowed concentrations of the caustic to a sufficient level on the tubes to produce the stress-corrosion cracking.[21, 22] Elimination of the free caustic from the water and installation of additional steam risers were the successful remedial measures.

At temperatures over 150°F, carbon and low-alloy steels become increasingly susceptible to stress-corrosion cracking by the caustic.

398 *Defects and Failures in Pressure Vessels and Piping*

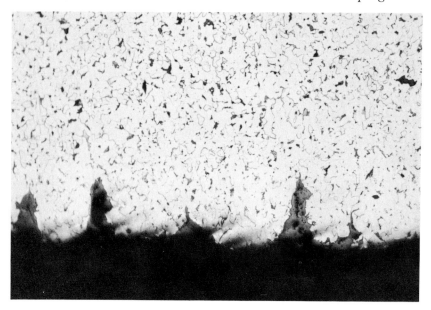

Fig. 17-28. Oxidation stress-corrosion involving cracking and oxygen corrosion pitting in boiler tubing. (Magnification 100×)

Heat accelerates the attack; hot spots where steam coils or other hot lines enter a vessel are particularly susceptible.[20]

Oxidation Stress Corrosion

Another type of stress corrosion occurs in the presence of oxygen. It has been principally reported in boiler tubing carrying feed and boiler water.[23-27] In the presence of tensile stresses, generally, intergranular cracking occurs. The cracks are widened by pitting corrosion as illustrated in Fig. 17-28.

CORROSION FATIGUE

In corrosion fatigue, the stresses that contribute to failure are associated with externally applied mechanical fatigue loads. Vibrations caused by a compressor in a piping system have been found a typical condition contributing to corrosion fatigue in a number of instances. Actually, the vibrations may not even be very severe. However, the resulting stresses act additively to the existing internal ten-

Fig. 17-29. Transgranular path of cracking in carbon steel pipe along slip planes of metal grains due to corrosion fatigue. (Magnification 500×)

sile stresses. Thus, the resulting mechanical fatigue may bring the stresses beyond the level at which cracking occurs.

In corrosion fatigue, the corrosion lowers the endurance limit of the steel (associated with the "roughening of the surface and, in severe cases, notch formation), and the fatigue accelerates the corrosive attack. In most environments where only corrosion occurs, the corrosive products and films that form tend to block or retard further corrosive attack; they may even cause it to cease completely. In corrosion fatigue, however, the cyclic stresses tend to rupture and render more permeable the protective areas.

Because of the mutually accelerating nature of corrosion and fatigue, a larger number of solutions and at lower concentrations can

400   *Defects and Failures in Pressure Vessels and Piping*

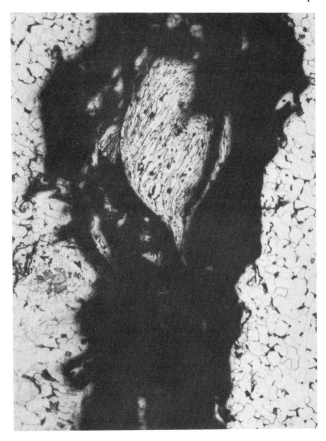

Fig. 17-30. Severe metal distortion caused by corrosion fatigue in carbon steel pipe. (Magnification 250×)

cause corrosion fatigue than can cause (static) stress-corrosion cracking.

Even though the stress level applied is below the fatigue limit characteristic of the metal, notches, cavities and other defects that act as local stress raisers support distortion (plastic deformation). In the absence of corrosion, the distortion (work hardening) in the structure raises the elastic limit in this area. In due course, the area subjected to an alternating stress of a given magnitude (below the fatigue limit) will resist further distortion because of the increase in the elastic limit. If, however, a corrosive agent is present, the distorted

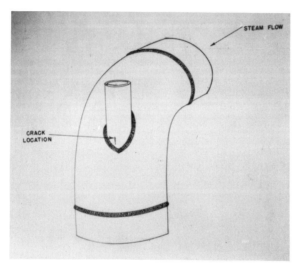

Fig. 17-31. Corrosion fatigue failure in 4-in. diameter pipe nozzle welded to 12-in. diameter steel elbow.

structure formed locally may become susceptible to corrosive attack. The potential for corrosion between distorted and undistorted structures can be significant, and the latter may remain unattacked. The combined chemical-mechanical action causes very fine cracks to advance inward, frequently along slip planes in a transgranular pattern as shown in Fig. 17-29. Severe distortion of the metallurgical structure in the affected area, Fig. 17-30, may be further evidence that corrosion fatigue has occurred.

(A)

(B)

Fig. 17-32. Progression of cracking in 4-in. diameter pipe nozzle as a result of corrosion fatigue. [Magnifications (A) 8× and (B) 15×]

Service Failures—Corrosion

Corrosion-fatigue failures are particularly likely to occur near restraints, attachment welds, nozzle welds, etc. A failure in a pipe nozzle weld is shown in Figs. 17-31 and 17-32.

Expansion bellows sections are also susceptible to corrosion-fatigue failures. An example of a corrosion-fatigue failure in a copper expansion bellows section in a hot water distribution system is illustrated in Fig. 17-33.

Elevated Temperature Service

At elevated temperatures, corrosion fatigue is generally enhanced by oxidation. The failure results from oxide formation and fatigue.

Corrosion-fatigue cracking as a result of sudden cold start-up has occurred in bends in risers and downcomers in boilers when the water contains a slight excess of oxygen.[28, 29] The resulting electrochemical attack reaches maxium intensity at about 300°F. Because of increasing scale protection at higher temperatures, it disappears at temperatures over 500°F.[28] The attack occurs at locations where insufficient protective Fe_3O_4 has formed on the tube wall, or has been damaged as a result of mechanical fatigue during start-up or shutdown. The initial cracking tends to occur along the neutral axis near the center wall portion of tube bends, and may not be visible on the outside and inside tube surfaces.[29] Subsequent failures by sudden rupture appear similar to the hydrogen embrittlement failure, illustrated in Fig. 16-20. Failures have not been observed in straight tube sections.[29]

In another instance, sudden rupture occurred in a $3\frac{1}{4}$-in. OD by 0.18-in. wall cold-bent downcomer. On the inside of the tube, 0.12-in. deep oxide-filled fatigue cracks were observed containing deposits of copper and copper oxides. Oxygen from the copper oxides was believed to have been a major factor in the oxidation of the steel and corrosion fatigue.[30]

The susceptibility of cold-bent downcomers to corrosion fatigue depends on the service condition. Where fatigue and superimposed stresses cause cracking of the protective oxide film on the tube inside, failure will occur. In such instances, hot-bent tubing or stress-relieved cold-bent tubing should be used.[30]

Susceptibility to these failures increases also with ovality. Failures have occurred in $2\frac{3}{4}$-in. diameter by 0.18-in. wall steel tubing pri-

404   *Defects and Failures in Pressure Vessels and Piping*

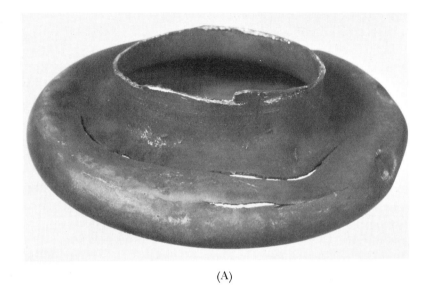

(A)

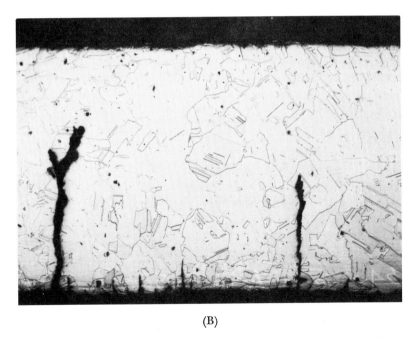

(B)

Fig. 17-33. Corrosion-fatigue failure in copper bellows section. [Magnifications (A) ½× and (B) 70×]

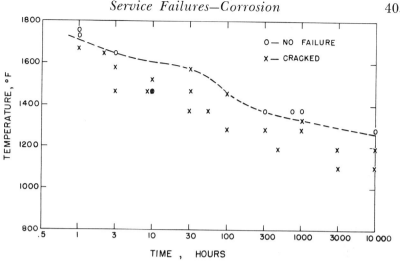

Fig. 17-34. Effects of stress relieving on resistance to stress corrosion in MgCl₂ solutions of electric-arc welded 16 Cr–13 Ni–Cb super-heater tubing and related to expected failures in superheater tubing due to corrosion fatigue.[32]

marily along the neutral axis of the cold-bent downcomers after 32,500 hours of service. The tubing contained ovalities of 12% producing high stresses along the neutral axis. Calculations showed the stresses to be approximately more than twice the yield strength. Thus, during every shutdown of the boiler, the protective scale on the tube inside was destroyed.[30]

Corrosion fatigue failures have been reported [31] also in 7Cr–½Mo vacuum flasher tubing in refinery service. The cracking occurred on the fire side of the tube near a restraint causing localized stressing in the tubing. The cracks starting from the surface were filled with oxides. During a hammer test, the tube failed in a brittle manner.[31]

Because of corrosion-fatigue cracking failures in 16Cr–13Ni–Cb superheater tubing, tests were conducted to determine suitable stress-relief heat treatments, Fig. 17-34. The test results in magnesium-chloride showed that even 10,000 hours at a service temperature of 1200°F do not reduce stresses sufficiently to eliminate susceptibility to stress-corrosion cracking.[32] As a result, stress-relieving temperatures close to 1920°F are recommended for superheater service applications.[32]

Remedies

Remedies for stress-corrosion cracking, caustic embrittlement, and corrosion fatigue include: 1) change in materials to a composition not susceptible to cracking in the particular environment; 2) change in water or chemical composition of the corrodent to minimize the elements promoting stress-corrosion cracking; 3) proper location of anchor points, hangers, and supports to minimize tensile stresses; 4) reduction of mechanical vibrations and service conditions that cause fatigue; 5) stress-relieving of weld joints or cold worked materials, consisting of slow heating and cooling uniformly around the circumference of the weld; and 6) changes in operating conditions which are known to have caused cracking during operating cycles, or shutdown, or start-up periods.

STRESS-ENHANCED CORROSION

Stress-enhanced corrosion may be the result of many different conditions. The corrosive attack takes place in areas containing high

Fig. 17-35. Stress-enhanced corrosion adjacent to weld seam as a result of the action of sulfuric acid and residual welding stresses in Type 316 stainless steel material.

Service Failures—Corrosion 407

residual stresses such as near welds, Fig. 17-35; machined surfaces, Fig. 17-36, or cold-worked or formed surfaces. The last condition was illustrated in Fig. 14-7 which showed corrosion by concentrated sulfuric acid along the outer arc of a seamless carbon steel pipe that had been cold bent. Since the corrosive environment was known in that particular case, the cold bending, rather than the service conditions, was considered the primary cause of failure.

In these illustrations, the corrosive attack tends to produce grooves or channels. The failure examples in Figs. 17-35 and 17-36 represent a change in operating conditions in a chemical plant. They occurred as a result of increased sulfuric acid formation in a vapor zone where sulfuric acid condensed on the Type 316 stainless steel surface.

A rather unusual example of this corrosion was observed in a Type 304L stainless steel piping system where the attack occurred in that part of the heat-affected zone normally susceptible to intergranular corrosion. This area will also retain the highest level of residual stresses after welding. As illustrated in Fig. 17-37, the corrosive at-

Fig. 17-36. Stress-enhanced corrosion in machined surface in Type 316 stainless steel caused by sulfuric acid.

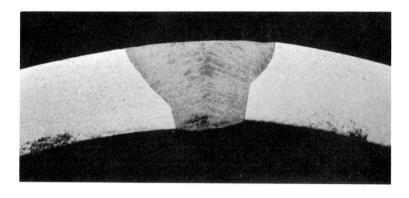

(A)

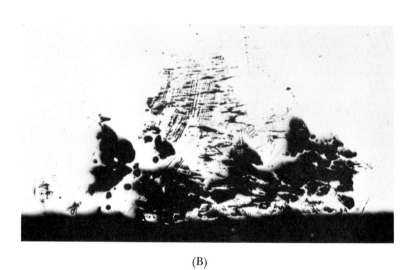

(B)

Fig. 17-37. Stress-enhanced corrosion in heat-affected zone of seam weld and in weld itself in Type 304L stainless steel pipe. [Magnifications (A) 5×, (B) 100×, and (C) 500×]

(C)

Fig. 17-37 *Cont'd*

tack progressed along slip planes. Intergranular attack was not observed. The chemical carried by the pipe was monovinyl chloride.

The proper remedies for stress-enhanced corrosion are either stress-relief heat treatments, or changes in the chemical solutions, or the materials specified. Heat treatments may not always be practical since they may produce excessive distortion.

EROSION

Erosion signifies the destruction of a metal by the abrasive action of a liquid or vapor. The presence of solid particles of matter in sus-

Fig. 17-38. Failure by erosion in carbon steel elbow caused by cavitation of steam in water.

pension, or entrained liquids in a vapor, may further accelerate this type of attack. In addition, erosion is often augmented by corrosion.

Erosion generally occurs in areas where the flow is restricted or the direction of flow is changed. Erosion may occur primarily in inlet and outlet nozzles, at vessel or pipe walls opposite inlet nozzles, on impingement baffles, in elbows, and in valves where flow is reduced or changed.

Where erosion is the result of the formation and collapse of cavities in a liquid at the solid-liquid interface, the corrosive attack is also known as cavitation. An example of this is shown in Fig. 17-38 illustrating the erosion produced by the cavitation of steam in water.

Service Failures—Corrosion 411

Fig. 17-39. Erosion caused by steam containing sawdust at edge of flange and weld between welding neck flange and 8-in. Schedule 40 carbon steel pipe.

Fig. 17-40. Erosion by wet steam in turbine at locations where protective surface film had been destroyed.

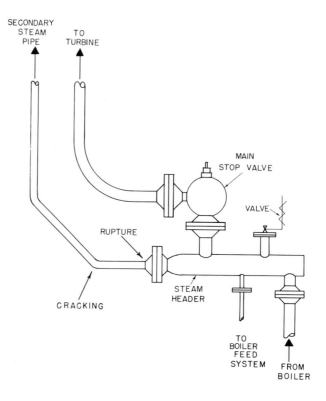

Fig. 17-41. Sketch illustrating failure location in 3½-in. OD by 0.18-in. wall steam pipe line.[30]

Attack ocurred only along the outer arc of carbon steel elbows, where the direction of liquid flow changed. Cavitation erosion has also been a major problem in hydraulic turbines and pumps.[33, 34]

A change in service conditions eliminating the occurrence of steam cavitation obviously is the most effective remedy for erosion. When this is not possible, a change to a material more resistant to attack, such as stainless steel, may also reduce or eliminate erosion.

Weld-overlaying critical areas subject to cavitation with wear resistant alloys is another effective method.

Erosion by steam is illustrated in Fig. 17-39. The steam eroded the edges of the flange joint and the weld metal in the welding neck flange to pipe joint.

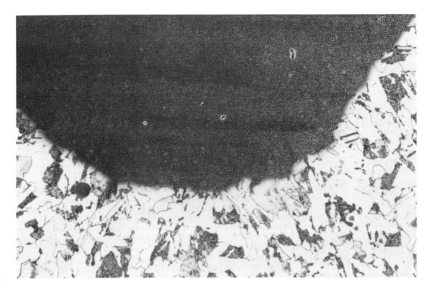

Fig. 17-42. Erosion of steel surface in exhaust end of top cover turbine shell caused by wet steam. (Magnification 100×)

Wet steam is also a frequent cause of erosion. Paint films or scale may inhibit erosion. However, where the protective surface film is destroyed, localized erosion may then occur, Fig. 17-40.

Erosion was also responsible for failure in a 45 degree elbow and rupture of the flange to pipe weld in a 3½-in. OD by 0.18-in. wall hot steam pipe line of ½Mo steel operating at 750°F and 400 psi. The cause was frequent introduction of boiler feed water into a steam header immediately ahead of the steam pipe due to an improperly operating valve, Fig. 17-41.[30]

Weld spatter and broken-off icicles remaining in piping after erection have also contributed to severe erosion.[35]

In refinery applications, the introduction of the fluid catalytic cracking process has led to severe erosion.[36] In this process, catalyst particles approximately 75μ in size circulate through the equipment like dust in a very dense dust storm. Temperatures vary from 900 to 1100°F. In some plants, the circulation rate is about 40 tons per minute. In the original units constructed in the 1940's, very rapid erosion occurred at bends, cyclone separators, valves and other locations where the flow was turned. Erosion has been significantly re-

414 *Defects and Failures in Pressure Vessels and Piping*

duced by improved design and the use of more erosion-resistant materials.[36]

Erosive attack of metals is generally characterized by wear across the metallurgical grains, as illustrated in Fig. 17-42. Selective intergranular-corrosive attack is normally absent.

EXFOLIATION

Exfoliation is a general type of corrosion which progresses essentially parallel to the surface. The corrosion products normally represent thick, leaf-type scale which peels away from the surface in flakes, Fig. 17-43.[37]

Exfoliation has occurred in 70-30 cupro-nickel alloys, and to lesser extents in 80-20 and even 90-10 cupro-nickel alloys. Common occurrences of exfoliation have been in heat-exchangers such as high-pressure feed water stage heaters in steam power plants.[37, 38] When the boiler operation involves peak load operations and reduced load

Fig. 17-43. Example of exfoliation attack on 70-30 cupro-nickel tubes of a high-pressure feed-water stage heater.[46]

operations and even shutdown periods, normal oxygen removal is not maintained. Thus, oxygen enters the system by inleaking. It is retained at significant levels until the system is returned to normal flow and pressures. During shutdown or low-load operation steam condenses, creating a vacuum. Although oxygen is considered the major element causing the exfoliation attack, ammonia and CO_2 are also believed to cause this attack.

The rate of exfoliation can be quite rapid. Tube failures in feed water stage heaters have occurred in only a few months of intermittent operation.

There has been some success in inhibiting this attack by blanketing the units with steam or inert gas immediately after shutdown. The steam or inert gas has to be maintained at a slight positive pressure (above atmospheric) to avoid inleakage of oxygen.

SOIL CORROSION

Almost all of the failures discussed above have generally resulted from corrosion starting on the inside of the vessel, tank or pipe. Unexpected service failures may also originate from the outside. The most common causes are associated with the insulation or covering placed around a vessel or pipe. In underground piping systems, these failures may be the result of soil conditions.

A failure in an underground pipe section is shown in Fig. 17-44. This type of attack is frequently associated with electrochemical reactions. Normally, this involves a difference in the electric potential between two points in contact with an electrolyte, where the electric current generated flows from the anodic area through the electrolyte to the cathodic area. Corrosion at the anode is caused by metal ions leaving this area. Among the various factors contributing to this type of attack is the moisture content of the soil and differential aeration.[39] The presence of oxygen, bacteria, and soluble salts also enhances attack.

Corrosion protection of underground steel piping normally calls for cathodic protection, usually with magnesium anodes or rectifier installations, or with tapes and coatings. Because of the many soil conditions encountered, no method of protection has ever been found satisfactory for all conditions and environments.

Fig. 17-44. Underground corrosion of 2-in. diameter carbon steel gas pipe in area where pipe insulation was damaged.

CORROSION CAUSED BY INSULATION

The pipe insulation employed may contain components that, under some conditions, contribute to corrosive attack. For example, water dripping on pipe insulation containing chloride or fluoride salts may convey them to the metal, causing corrosion of the piping materials. This is illustrated in Fig. 17-45.

Leaching of magnesium chloride salt from the insulating wall refractory during shutdown periods has resulted in stress-corrosion cracking starting from the outside of Type 347 stainless steel tubing.[40]

SCALING AND OXIDATION

Corrosion involving oxidation and scaling on the outside surfaces of tubing at high temperatures is of great interest in steam boilers. This subject, covered very extensively in the literature, is not treated here.[41-44]

Service Failures—Corrosion

Fig. 17-45. Corrosion of pipe surface caused by water dripping on insulation around pipe.

ANALYSIS OF FAILURES

The corrosion failures discussed in this Chapter by no means exhaust the types that can occur with excessively severe service conditions. Frequently, it is quite difficult to diagnose the basic cause in a failure by corrosion. This is true when operating records are not kept, or when an improper fluid has been pumped through the systems.

A current problem inherent to radio-chemical processing is the difficulty of predicting a metal's or alloy's corrosion resistance to radiochemicals having high radiation levels. The combination of aggressive fluids and radioactive elements or compounds in some services precludes the possibility of precisely predicting corrosion. This results from the relatively unknown corrosive effect that the radioactive elements may confer on the solution of which they are a part. Laboratory work with such solutions is necessarily severely restricted. Further, metallic components that failed in such an environment may be so intensely contaminated by radioactive materials that a complete examination is impossible. Such components must then be conveyed to disposal by remote control, without the examination that might make it possible to prevent a recurrence of the failure. It should be noted, however, that decontamination capabilities are far superior to what they were in the past, and there is promise of further advances in this field.

If a corrosive failure analysis does not result in obvious answers, it may be advisable to immerse special test sections of different materials in the actual service environment. These sections must be carefully prepared so as not to lead to erroneous conclusions.

REFERENCES

1. Uhlig, H. H. (Editor), "The Corrosion Handbook," John Wiley & Sons, New York, 1948.
2. La Que, F. L. and Copson, H. R., "Corrosion Resistance of Metals and Alloys," Second Edition, Reinhold, New York, 1963.
3. Jacklin, C., "Waterside Failures in High-Pressure Boilers—A Field Survey," *Am. Soc. Mech. Eng.*, Paper No. 61-WA-271 (1961).
4. Krebs, T. M., "On-the-Site Conditions Affect Corrosion," *Oil Gas J.*, **55**, 140 (Feb. 4, 1957).
5. Krebs, T. M., "Pitting Corrosion Attack on Steel," *Oil Gas J.*, **55**, 125 (Mar. 4, 1957).
6. Resch, G. and Odenthal, H., "Die Korrosion des Eisens in sauerstoffhaltigem Kondensat," *Mitt. Ver. Grosskesselbesitzer*, **76**, 4–11 (1962).
7. Kaes, H., "Stillstandskorrosionen in Vorwärmern," *Mitt. Ver. Grosskesselbesitzer*, **82**, 62–64 (1963).
8. Corey, R. C., "Corrosion of High-Pressure Steam Generators: Status of Our Knowledge of the Effect of Copper and Iron Oxide Deposits in Steam Generating Tubes," *Proc. ASTM*, **48**, 907–939 (1948).
9. Peters, B. F., "Radiography for Corrosion Evaluation," British Columbia Metals Conference, June 5, 1964.
10. Robertson, W. D., "Stress Corrosion Cracking and Embrittlement," John Wiley & Sons, New York, 1956.
11. *Symposium on Stress-Corrosion Cracking of Metals,* Am. Soc. Testing Mater., Philadelphia (1945).
12. Copson, H. R., "The Influence of Corrosion on the Cracking of Pressure Vessels," *Welding J.*, **32**, Res. Suppl., 75s–91s (1953).
13. Gall, W. R., "Heavy-Water Leaks at Savannah River," *Nucl. Safety*, **3**, 87–90 (Sept. 1961).

14. Griess, J. C., "Stress Corrosion Cracking of Stainless Steel," *Nucl. Safety,* 4, 32–35 (Mar. 1963).
15. Schaffer, L. D. and Klapper, J. A., "Investigation of the Effects of Wet, Chloride-Bearing, Thermal Insulation on Austenitic Steel Pipe," USAEC Report ORNL-TM-14, Oak Ridge National Laboratory (Nov. 1, 1961).
16. Zapffe, C. A., "Boiler Embrittlement," *Trans. ASME,* 66, 81–126 (1944).
17. Weir, C. D., "Recent Research on Caustic Cracking Boilers," *Trans. Inst. Engrs. Shipbuilder Scot.,* 92, 165–191 (Jan. 1949).
18. Martels, W., "Methodik bei der Untersuchung von Schäden an geschweissten Bauteilen," *Schweissen und Schneiden,* 8, 269–277 (1956).
19. Sines, G. and McLean, E. C., "The Failure of a Welded Drying Drum by Caustic Embrittlement," *Mech. Eng.,* 78, 1105–1109 (1956).
20. Guide for Inspection of Refinery Equipment, VI Unfired Pressure Vessels, API Division of Refining, New York (1958).
21. Singley, W. J., et. al., "Stress Corrosion of Stainless Steel and Boiler Water Treatment at Shippingport Atomic Power Station," *Proc. American Power Conference,* 21, 748–766 (1959).
22. McCurdy, H. C. and Anderson, T. D., "Shippingport Operating Experience," *Nucl. Safety,* 2, 102–107 (Sept. 1960).
23. Rädeker, W., "Übergangsmöglichkeit zwischen Spannungsrisskorrosion und Lochfrasskorrosion bei weichen Stählen," *Arch. Eisenhüttenw.,* 30, 745–750 (1959).
24. Rädeker, W., "Die Spannungsrisskorrosion an niedriglegierten Stählen für den Dampfkesselbau," *Mitt. Ver. Grosskesselbesitzer,* 74, 332–344 (1961).
25. Bachmair, A. and Kaes, H., "Risschäden an unbeheizten Verbindungsrohren," *Mitt. Ver. Grosskesselbesitzer,* 69, 424–429 (1960).
26. Bachmair, A. and Kaes, H., "Ausgewählte Korrosionsfälle aus dem Kraftwerksbetrieb," *Mitt. Ver. Grosskesselbesitzer,* 75, 411–421 (1961).
27. Jahn, E. and Deeskow, M., "Rohrschäden unter gleichzeitigem Einfluss von Spannung und Sauerstoff," *Mitt. Ver. Grosskesselbesitzer,* 79, 294–300 (1962).
28. Kiekenberg, H., "Einfluss der Ausführung von Vorwärmern auf den Eisenspiegel im Wasser-Dampf-Kreislauf," *Mitt. Ver. Grosskesselbesitzer,* 85, 232–239 (1963).
29. Schneider, A., "Vermeidung von Schäden beim An- und Abfahren von Kesseln und Blockanlagen," *Mitt. Ver. Grosskesselbesitzer,* 86, 271–287 (1963).
30. Verein. d. Grosskesselbesitzer, 1962/63 Review.
31. Krebs, T. M., "Intergranular Cracking and Corrosion," *Oil Gas J.,* 55, 157 (May 27, 1957).
32. Ruttmann, W. and Brunzel, N., "10 Jahre austenitische Stähle im Kesselbetrieb," *Mitt. Ver. Grosskesselbesitzer,* 80, 310–326 (1962).
33. Knapp, R. T., "Accelerated Field Tests of Cavitation Intensity," *Trans. ASME,* 80, 91–102 (1958).
34. Kerr, S. L. and Rosenberg, K., "An Index of Cavitation Erosion by Means of Radioisotopes," *Trans. ASME,* 80, 1308–1314 (1958).
35. Verein. d. Grosskesselbesitzer, 1959/60 Review.
36. Finnie, I., "Erosion by Solid Particles in a Fluid Stream," *Symposium on Erosion and Cavitation,* 70–82, ASTM Tech. Publ. No. 307 (1962).

37. Wilkes, J. F., "Corrosion and Embrittlement Protection for Boiler Systems," *Materials Protection,* **1,** 18–26 (May 1962).
38. Neat, F. U., "The Relation Between Boiler Cleanliness and Feedwater," *Combustion,* **32,** 35–42 (Mar. 1961).
39. Schaschl, E. and Marsh, G. A., "Some New Views on Soil Corrosion," *Materials Protection,* **2,** 8–17 (Nov. 1963).
40. Krebs, T. M., "How To Prevent Stress Corrosion Cracking," *Oil Gas J.,* **55,** 207 (May 20, 1957).
41. Jackson, P. J. and Ward, J. M., "Operational Studies of the Relationship Between Coal Constituents and Boiler Fouling," *J. Inst. Fuel,* **29,** 154–164 (1956).
42. Jackson, P. J. and Raask, E., "A Probe for Studying the Deposition of Solid Material from Flue Gas at High Temperatures," *J. Inst. Fuel,* **34,** 275–280 (1961).
43. Alexander, P. A., Fielder, R. S., Jackson, P. J., Raask, E., and Williams, T. B., "Acid Deposition in Oil-fired Boilers: Comparative Trials of Additives and Testing Techniques," *J. Inst. Fuel,* **34,** 53–72 (1961).
44. Jackson P. J., "Feuerseitige Ablagerungen und Korrosionen in Dampferzeugern," *Mitt. Ver. Grosskesselbesitzer,* **85,** 220–231 (1963).
45. Straub, F. G., "Embrittlement in Boilers," University of Illinois, Engineering Experiment Station Bulletin No. 216 (1930).
46. Tucker, R. C., Private Communication, Niagara Mohawk Power Corp., Oswego, N. Y.
47. Pearson, C. E. and Parkins, R. N., "Stress-Corrosion Cracking in Welded Steel Structures," *Trans. Inst. Welding,* **12,** 95r–106r (1949).

chapter 18

MEANINGFUL INTERPRETATION OF INSPECTION RESULTS*

In recent years, much has been said about unrealistic and excessively tight quality assurance requirements applied to nuclear power plants that have increased plant costs by many millions of dollars without contributing to greater plant system integrity. Increasingly tight requirements are being applied to the materials utilized in piping systems, valves, vessels, containment and other components in nuclear power plants.

Many unrealistic rejections are related to a lack of understanding of engineering materials and their performance. Thus, understanding the characteristics of engineering materials and their performance in different shapes in specific service environments is essential to more realistic quality assurance programs.

Responsible materials engineering includes recognition that although plants must be safe and reliable and must utilize materials and products that meet code requirements, they must be engineered and constructed at the lowest possible cost.

The overall purpose of quality assurance is to avoid service failures. Failures, obviously, may have been caused by defects in materials,

---

*This chapter is based on a paper presented at the International Energy Engineering Congress at Chicago, Ill. on November 5, 1975 under the sponsorship of James O. Rice Associates, Inc., 400 Madison Ave., New York, N.Y. 10017.

422   *Defects and Failures in Pressure Vessels and Piping*

Fig. 18-1. Rupture of 12 in. outside diameter stainless steel elbow due to cracking along weld overlaid heat affected zone adjacent to seam weld.

products, or design, but failures have also been caused by excessive rework to provide "cosmetic" improved appearance.

Unrealistic or misunderstood quality assurance may result in excessive repairs of engineering materials. Such repairs, although improving the "interpretability" of inspection results, may in fact reduce the integrity of the repaired components. A severe catastrophic failure, for example, was caused by excessive cosmetic grinding of a stainless steel fitting. The ruptured seam welded elbow involved is shown in Fig. 18-1. Upon completion of the seam weld in the 14 in. stainless steel elbow, the manufacturer ground the weld seam. Because of overgrinding, which reduced the base metal wall thickness excessively, he subsequently applied a weld overlay extending by 1 in. over the base metal beyond the original butt weld. The particular stainless steel involved is subject to severe grain growth in the heat affected zone, which exhibits lower high strength properties at elevated temperature than either base metal or the weld metal. The cracking that occurred in the heat affected zone from the inside

Meaningful Interpretation of Inspection Results 423

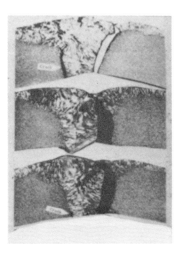

Fig. 18-2. Cross section illustrating crack progression through heat affected zone and detailing weld overlay along outside surface.

Fig. 18-3

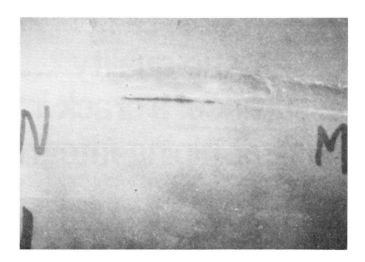

Fig. 18-3. Cracking of 18 in. outside diameter stainless steel pipe through heat affected zone along edge of circumferential butt weld that had not been weld overlaid.

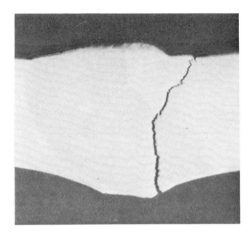

Fig. 18-4. Cross section illustrating cracking through heat affected zone of normal weld without weld overlay on pipe surface (50 X magnification).

surface progresses through the pipe wall to the weld overlay. Because of the greater high temperature strength of the weld metal, the crack then progressed longitudinally underneath the overlay for a distance of about 8 in. At this stage, the catastrophic rupture occurred, which also resulted in extensive plant damage. In the absence of the weld overlay, the crack would have continued to progress through the pipe wall adjacent to the seam weld (Fig. 18-2). At some stage, the surface then would have opened by cracking for about ½ to 1½ in. (Figs. 18-3 and 18-4). This would have resulted in steam leakage and provided a warning to the operator. Repair or replacement of the elbow could have been readily accomplished. The cost to the plant operator would then have been measured in thousands of dollars rather than millions.

Many similar examples of failures can be cited that can be related to needless manufacturing operations. Fortunately, these generally have not resulted in catastrophic consequences, even though they may have been very costly.

Most pipes and forgings contain superficial surface laps, slivers, tears etc. These are recognized in many of the ASME (American Society of Mechanical Engineers) materials specifications as acceptable within specific limits. An example of surface laps in carbon steel pipe for a feedwater system in a nuclear plant is shown in Fig. 18-5. A typical lot of pipe illustrated contained 90 percent surface laps penetrating into the pipe wall to a depth of less then 5 percent, 9 percent involving laps at depths between 5 and 12.5 percent of the wall thickness, and an occasional lap involving depths of approximately 15 percent of the wall thickness. On one nuclear power plant project, all of those indications were ground out in a number of lengths of carbon steel pipe. The grooves were subsequently checked for wall thickness ultrasonically. Since the grinding of the tight superficial surface laps was done to significantly greater depths than the actual depths of the laps, *i.e.,* deeper than 12.5 percent of the

Fig. 18-5. Illustration of typical surface lap on 14 in. seamless carbon steel pipe.

426   *Defects and Failures in Pressure Vessels and Piping*

A

Fig. 18-6. Photo A shows surface laps on seamless 20 in. outside diameter Schedule 80 (1.031 in. wall) carbon steel pipe from 0.023 to 0.056 in. deep representing 2.2 to 4.9 percent of the pipe wall thickness. Photo B shows surface lap on seamless 26 in. outside diameter by 0.975 wall carbon steel pipe 0.075 in. deep representing 8.6 percent of pipe wall thickness.

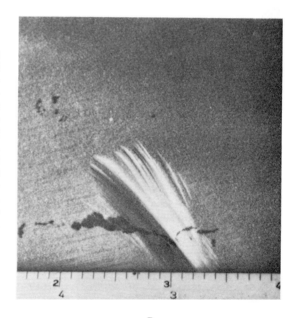

B

nominal wall thickness, considerable rewelding was necessary at the nuclear power plant construction site. Major schedule delays and high costs resulted. Grinding or filing across a few selected surface laps was an entirely acceptable procedure that confirmed that the laps in the particular piping were completely inconsequential (Fig. 18-6). The specific manufacturer's pipe products were thus proven entirely acceptable.

Many similar resolutions applied to various other piping products can be cited. Additional examples in stainless steel pipe, forged fittings, and forged valves, where surface defects have been confirmed as acceptable by material and mechanical test analyses, are shown in Figs. 18-7 to 18-9

Whereas piping material specifications broadly define welding processes, variations in the processes and/or procedures can result in defects not readily detectable. On the average, pipe and fitting seam welds welded without filler metal tend to be superior to seam welds made with filler metal. The reason is that for welding seams in piping and fittings without filler metal, the inert gas tungsten arc welding processes must be employed. Thus, less rigid inspection requirements should apply to fittings and pipe seams welded without filler metal.

With respect to shop fabrication or field erection, welding procedures generally are pre-established and written with rather tight control over various welding parameters. When specific problems then arise during production welding, it often becomes extremely difficult or at times impractical or even impossible to make changes in specific welding procedures or parameter details. For example, on one nuclear project, the welding or containment pipe sections involved both welds in 24 in. diameter pipe located 12 in. from the containment wall. The pipe had been prepared with J-bevel preparation involving a 10 deg taper on the J-bevel. The welders had problems with electrode manipulation because of interference from the containment wall (Fig. 18-10). The weld deposits were unevenly filled; *i.e.,* each weld layer was an angle rather than parallel to the centerline (Fig. 18-11). The welding procedure also contained a weave width of three times the electrode core diameter. The problem with electrode manipulation resulted in significant slag entrapment along the bevel side where the welder was located. Radiographic examination resulted in weld rejections of over 50 percent due to excessive slag

428 *Defects and Failures in Pressure Vessels and Piping*

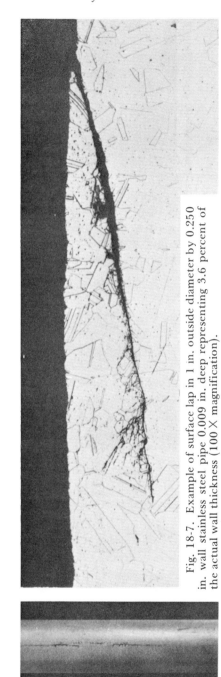

Fig. 18-7. Example of surface lap in 1 in. outside diameter by 0.250 in. wall stainless steel pipe 0.009 in. deep representing 3.6 percent of the actual wall thickness (100 × magnification).

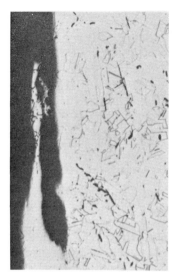

Fig. 18-9. Example of forging laps on 1 in. forged Type 304 stainless steel elbow confirmed as non-injurious (100 × magnification).

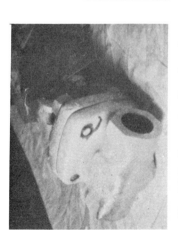

Fig. 18-8. Example of surface laps at bolt hole of 1 in. forged stainless steel valve.

Meaningful Interpretation of Inspection Results 429

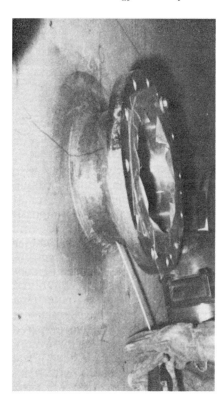

Fig. 18-10. View of butt joint in 20 in. outside diameter pipe adjacent to penetration illustrating difficulty in manipulating electrode perpendicular to pipe butt joint groove.

Fig. 18-11. Close-up view of deposited weld layer illustrating uneven weld level in groove.

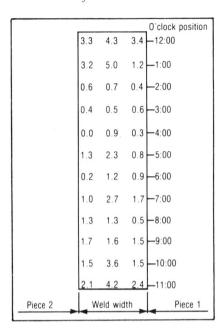

Fig. 18-12. Delta ferrite measurements around circumference of 2 in. wide butt weld in 14 in. outside diameter by 1.25 in. wall Type 316 stainless steel pipe.

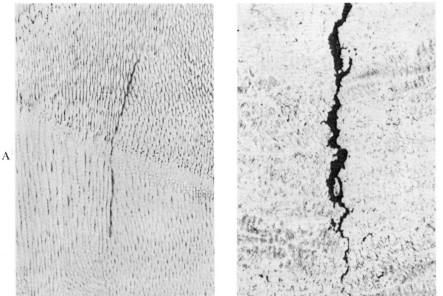

Fig. 18-13. Examples of microfissures in Type 316 stainless steel weld joints containing 1 percent delta ferrite (Photo A, left) and 10 percent delta ferrite (Photo B, right). Both photos are reproduced at 100 X magnification.

Meaningful Interpretation of Inspection Results 431

inclusion. An immediate change in the field welding procedures, involving grinding the weld bevel on the pipe side of the joint to a 45 deg V-bevel and increasing the weave width to 1 in., reduced rejections because of slag inclusions to less than 5 percent. In accordance with ASME Section III and 10 CFR 50 requirements, records, of course, were kept of the change. The procedure modification was approved at the field erection site by the responsible personnel from welding engineering, QA/QC (quality assurance/quality control), and management.

On most nuclear power plant projects, changes in welding procedures involve such a cumbersome review and approval cycle, with many reviewers located far from the project site adding unrelated changes, that it takes months or years before welding procedures can be changed.

In addition to poorer weld quality, excessive weld repairs, schedule delays, and greatly increased costs are results of excessively complicated or unrealistic requirements applicable to fabrication, construction, or erection procedures.

Rather tight requirements have been applied to weld deposits in stainless steel piping with respect to delta ferrite levels. Many welds have been rejected and have required rewelding because they contained delta ferrite levels of less than 5 percent or greater than 15 percent. Extensive delta ferrite measurements from various nuclear power plant projects showed levels as low as ½ percent or even as high as 25 percent or higher. An example of low ferrite levels in an entirely acceptable weld in Class 1 piping is shown in Fig. 18-12. Experience with stainless steel welds in the chemical industry also confirms that fully austenitic weld deposits have not failed in service, even though microfissures were present. In fact, microfissures can be present in welds with delta ferrite contents of 10 to 15 percent (Fig. 18-13). The weld deposits nevertheless exhibit high levels of integrity.

Delta ferrite measurements made on stainless steel casting installed in many nuclear power plants have shown levels ranging from 0 to as high as 50 percent. These castings provide satisfactory service in nuclear power plant installations, even though castings generally contain greater levels of fissures and shrinkage cavities than weld deposits.

Stainless steel materials are also rejected unrealistically because of sensitization. With the conventional Type 304 and Type 316 stainless

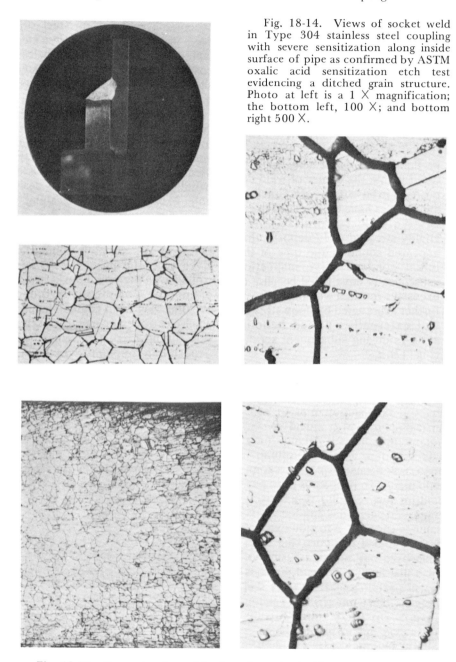

Fig. 18-14. Views of socket weld in Type 304 stainless steel coupling with severe sensitization along inside surface of pipe as confirmed by ASTM oxalic acid sensitization etch test evidencing a ditched grain structure. Photo at left is a 1 X magnification; the bottom left, 100 X; and bottom right 500 X.

Fig. 18-15. Example of sensitization along the inside surface of pipe sensitized by normal pipe forming. Photo at left is at 25 X magnification; photo at right is at 500 X magnification.

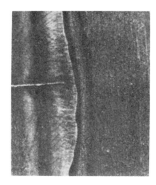

Fig. 18-16. Example of cracking in weld area caused by excessive weld repairs.

steel materials utilized in vessels and piping, welding will result in sensitization of the base materials. This cannot be avoided. A typical example of a Type 304 stainless steel coupling sensitized by welding is shown in Fig. 18-14. Hot forming, and even some heat treatment operations, have also caused severe sensitization in some stainless steel components installed in nuclear power plants (Fig. 18-15). Nevertheless, these materials have provided and continue to provide satisfactory service in BWR (boiling water reactor) and PWR (pressurized water reactor) nuclear power plants.

Although greater heat input during welding will increase the tendency toward sensitization, it will reduce excessive weld shrinkage and residual stresses. Excessive weld shrinkage and residual welding stresses in many instances are generally far more critical than weld defects of borderline levels—particularly when the defects are located within the center of weld deposits or in base materials. Cracking and failures have been related to excessive weld repairs. (Fig. 18-16). Highly stressed weld deposits that appeared sound at the time of the initial nondestructive inspection subsequently have failed because of excessive residual welding stresses, further aggravated by hoop stresses and externally applied bending stresses during subsequent service (Fig. 18-17).

Significant construction costs are incurred by requiring the insertion of stainless steel shims or strips between stainless steel pipe and the large number of carbon steel pipe supports, clamps, guides, channels, etc., used in the support of stainless steel piping systems in nuclear power plants (Figs. 18-18 and 18-19). Direct attachment of these supports to stainless steel pipe is entirely acceptable, as confirmed by the extensive satisfactory performance of such sup-

Fig. 18-17. Example of leaks through socket weld primarily related to residual welding stresses.

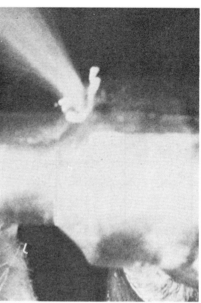

Fig. 18-18. Example of shimming with stainless steel strip metal connection between carbon steel pipe clamp and stainless steel pipe.

ports in existing nuclear power plants (Fig. 18-20) and fossil fuel power plants. Even in outdoor chemical plants in coastal areas, where rains or hurricanes have rained salt water on such connections, the stainless steel pipe has not deteriorated (Fig. 18-21). Laboratory corrosion test data also confirms the acceptability of carbon steel supports in contact with stainless steel piping and vessels. Nevertheless, costly shimming with stainless steel sheet metal is done on many nuclear power plant projects.

Another example of needless costly rework involves the concern over evidence of surface rust in stainless steel materials. Although cleanliness in the fabrication or handling of stainless steels is desirable, excessive surface cleaning to remove every spot of rust does not result in greater piping integrity. Even under the most perfectly controlled welding conditions, rusting cannot be completely avoided. Welding slag from coated stainless steel electrodes or metal varporization during inert gas welding will deposit thin iron films over the surface that tends to rust in the presence of moisture (Figs. 18-22 and 18-23).

Other examples of resolutions performed on behalf of various nuclear projects have included:

- Resolutions of laminations in containment plate materials and acceptable levels applicable to weld defects caused by laminar plate inclusions.
- Acceptance of laminations in pipe and plate materials detected by ultrasonic examinations.
- Mismatch and high-low acceptance considerations of butt welds to avoid needless counterboring, fit-up, and inspection problems.
- Welding procedures and weaving width changes to reduce slag inclusions and porosity and to minimize weld rejections and repairs.
- Resolutions and acceptance of various "contaminations" apparent on stainless steel weld, pipe, and vessel surfaces, and including discoloration, rust, chemicals, etc.
- Reducing gas purging requirements without affecting weld quality of pipe butt welds.
- Resolutions of weld rejections involving "excessive" weld reinforcement on the underside of weld joints as estimated by radiographic or ultrasonic examinations.
- Review resolution and acceptance of Class 3 or Class 2 piping installed in Class 2 or Class 1 piping systems.

436 *Defects and Failures in Pressure Vessels and Piping*

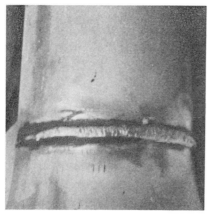

Fig. 18-22. Example of rust surface layer on stainless steel pipe adjacent to weld deposit made by shielded metal arc welding.

Fig. 18-23. Thin surface layer on stainless steel pipe adjacent to lug weld made by inert gas tungsten arc welding.

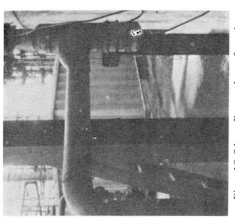

Fig. 18-21. Examples of carbon steel U-bolts on stainless steel pipe involving outdoor installation in chemical plant.

Fig. 18-19. Another illustration of shimming of stainless steel pipe with stainless steel strip metal in carbon steel support assembly.

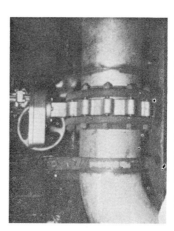

Fig. 18-20. Carbon steel pipe clamp and carbon steel slip-on flanges on stainless steel pipe in fossil fuel power plant.

- Acceptance of carbon steel to chromium-molybdenum dissimilar weld joints that were welded without preheat treatment.
- Acceptance of valves with wall thickness areas below minimum design wall thickness requirements, as determined by ultrasonic examinations, and/or actual measurements.
- Examination and resolution of piping materials, pumps, valves, and/or pipe supports subject to fire exposure.
- Evaluation, resolution, and acceptance of linear indications in weld deposits, weld repairs, or weld cladding overlays detected by liquid penetrant examination after surface grinding.

Many more examples could be cited!

By realistically evaluating the characteristics and performance of the materials used in nuclear power plants and applying these results in the specification and interpretation of quality assurance requirements, very significant cost reductions can be realized on many current nuclear power plant projects without in any manner adversely affecting the integrity of the piping, vessels, components, and systems.

Materials engineering know-how and extensive performance experiences of the respective components in various environments are essential ingredients in the resolution process. Much can be learned in this respect from product performance and failure analysis experiences in the chemical and petrochemical plants, fossil fuel power plants, paper mills, and mining and metal processing plants.

Total support of this by the utility's management is essential. If effectively implemented, the result would be reliable nuclear power plants at lower costs.

Similar reduction in costs and improvements in quality and integrity have been identified with respect to repair and maintenance, and in-service inspection.

Index

Acid cleaning, 362
Air compressor, 20
Aluminum alloys, 121, 197, 232, 243, 249, 256, 263, 375, 382
Ammonia storage vessel, 148
Arc strikes, 217-222, 295-296
Attachment welds, 29, 310, 390

Backing rings, 35, 208-210, 222, 261, 382
Backup flux, 286
Baffles, 32, 34
Bar stock defects, 113
Bark, 107
Blow holes, 98, 107, 131
Boilers, 292, 340, 394
Bond, 76
Brittle grain boundaries, 117
Brittleness, 4, 155
Burn through, 223
Butane storage tanks, 365
Buttering, 121

Carbon migration, 76
Carburization, 123
Casting defects, 130-146
CAT, 165
Catalytic hydroformers, 5, 284, 365
Cauliflower-type oxidation, 285
Caustic embrittlement, 295, 317, 394-398
Cavitation, 410
Centerline cracking, 230, 277
Centerline crevice, 226
Chaplet, 139
Chemical composition
  effects on cold cracking, 234
  effects on hot cracking, 229-233
  effects on notch brittleness, 176

Chills in castings, 140
Chromium stainless steels, 71, 197
Clad steels, 25, 353, 392
Cleaning, 325-332
Cleavage fractures, 155
Code limits on impact test requirements, 172-173
Code limits on surface defects, 111
Cohesive strength, 117
Cold bending, 195, 310, 312, 403, 407
Cold cracks, 233-234
Cold shock, 341, 347
Columbium in stainless steels, 67, 215, 287-289
Concavity, 223, 240, 250-253
Concentration-cell corrosion, 379
Condensate carry-over, 342-343, 391
Condensate pitting, 379
Copper alloys, 123, 232, 246, 252, 293, 387, 403, 414
Copper deposits, 119, 331, 373, 380, 403
Copper spectator deposits, 380
Corrosion, general, 371, 372
Corrosion fatigue, 311, 398-406
Corrosion pitting, 379-385
Crack arrest temperature (CAT), 165
Crack detection, 274
Crack growth, 10
Crack initiation, 9, 155, 157, 166
Crack propagation, 10, 155, 157, 341
Crack starter weld, 168
Cracks
  castings, 137
  welds, 39, 149, 155, 208, 226-238, 274
  wrought pipe, 109
Crater pits, 238-239, 279
Crater-type corrosion, 382
Crevice corrosion, 385-387

439

440 Index

Decarburization, 123, 140, 197, 284, 309, 358, 363
Defect
  definition, 1
  metallurgical, 9
  structural, 9
Dendrites, 131, 229
Dents, 85
Deoxidation of steel, 57, 64, 152
Design notch, 17, 185
Desuperheater piping, 346-348
Dilution, 232
Dissimilar metal chills, 140
Dissimilar metal joints, 73-80, 282, 289, 312
Distortion, 317
Drop weight impact test, 168
Dross, 131
Ductility, 4
Ductility transition, 157, 174

Elbows, 105, 123, 413
Electrode storage, 87
Embrittlement
  graphite, 53-64
  hot-shortness, 68, 90, 107, 117-123, 137, 142
  hydrogen, 355-364
  neutrons, 184
  notch, 147-193
  sigma, 71-72
  sulfur, 119
Energy transition, 157
Erosion, 32, 34, 210, 409-414
Esso test, 170
Eutectic structure, 117
Excessive stresses, 335
Exfoliation, 414-415
External loading, 336
Extruded outlets, 201

Failure mechanism, 8-11
Failures
  catastrophic, 5, 31, 148-151, 166, 190, 299, 339, 340, 351, 355, 392, 394, 395
  definition, 1
Fatigue limit, 337
Fatigue strength, 337

Feedwater heaters, 311
Ferrite in stainless steel welds, 231, 287
Fin, 104
Fires caused by failures, 5, 315
Fissures, 107, 121, 227
Fit-up, 207, 290, 382
Flame straightening, 316-319
Flanges, 20, 113, 335, 413
Flash welding, 279
Flexibility, 12-17
Forging defects, 113
Fracture analysis diagrams, 164-167
Fracture appearance transition temperature, 160
Fracture transition, 157, 174
Fracture transition elastic (FTE), 165
Fracture transition plastic (FTP), 165
Fretting corrosion, 210
Friction oxidation, 210

Galvanic corrosion, 373-376
Gas compressor piping, 337, 340
Gas shielded consumable metal arc welding, 277-279, 297
Gas transmission piping, 85, 116, 128, 161, 198, 250, 395
General corrosion, 371, 372
Gouges, 85, 114, 116
Grain boundary segregations, 142
Grain size, 178
Grapes, 223
Graphitization, 53-64
Grinding, 329

"**H**" superheater tube grades, 308
Hammering, 85, 150
Hanger lugs, 29, 339
Hardness variations, 31, 72, 116, 128, 142, 197, 219, 296, 304, 305-306, 311, 312, 317, 353
Headers, 13, 89, 392, 413
Heat exchanger failures, 293, 312, 315, 377, 393, 394, 397, 414
Heat shock, 341
Heat treating methods, 319, 322
Heat treatment, 301-324
  effects on notch brittleness, 181-183
Heating coils, 32
Hi-Low mismatch, 239-241

Hoop stress, 26, 149
Hot cracks, 229-233, 235
Hot ductility, 67, 117, 201, 216
Hot shortness, 68, 90, 107, 117-123, 137, 142
Hot tears, 137
Hot tensile tests, 120
Hydrogen blistering, 365
Hydrogen damage, 355-364
Hydrogen embrittlement, 284, 355-364
Hydrogen storage vessels, 148

Icicles, 223, 413
Impact energy absorption transition temperature, 160
Impact properties, 155
Impact tests, 168
Impurities, 117
Inclusions, 98, 131, 297
Incomplete fusion, 42, 241-242, 279
Inconel, 78, 233
Incorrect identification, 88-91
Induction heating, 319
Inert-gas tungsten-arc welding, 251, 255, 256, 273-277, 284-286
Insulation damage, 394, 416
Intergranular carbide precipitation, 66
Intergranular corrosion, 66, 307, 326, 377
Intermetallic phase, 121
Isothermal crack-arrest temperature, 167

Knife-line corrosion, 377-379

Laboratory tests, 2, 50, 64, 67, 174, 215
Lack of penetration, 40, 208, 242-245, 285, 292-295, 390
Laminations, 98-100, 297, 365
Laps, 2, 100-104
Liquation, 68, 117, 229
Liquefied gas containers, 148, 365
Low-cycle fatigue, 29, 337

Material selection, 49-83, 289
Mechanical fatigue, 23, 29, 337-339, 398
Mechanical notches, 97-117, 351
Mechanical shock, 339-340
Metallurgical defects, 9
Metallurgical homogeneity, 177

Metallurgical notches, 96, 117-128, 351
Metallurgical stabilization heat treatment, 63
Methane blisters, 284
Methane formation, 359-360
Methane vessels, 148
Microfissures, 119, 215, 227, 234, 237, 284, 287
Missile components, 215
Miters, 17, 286
Monel, 89

Nelson chart, 364
Neutron radiation, 184
Nickel alloys, 233
Nil ductility transition (NDT), 164-168, 175
No-break temperature, 169
Notch-brittle steels, 31, 147-193
Notch conditions, 38, 84, 116, 153, 163, 209
Notched tensile tests, 168
Nozzle joints, 23, 26, 43, 326, 403, 410
Nuclear reactor plant equipment and piping failures, 6, 89, 92, 184, 288, 307, 326, 340, 348, 353, 385, 393, 394, 397

Oil storage tanks, 149
Oil-well casings, 84, 104, 114, 128, 296
Ovality, 84, 200, 311, 403
Overfill, 104
Overheating, 352-355
Overlap, 245
Overpressure, 335
Oxidation stress corrosion, 398
Oxidation of weld surfaces, 245, 284
Oxyacetylene welding, 250

Penstock failures, 299, 339
Pickling, 325-327, 357, 363
Pipe bends, 17, 29, 120, 195-200, 310, 312, 403
Pipe supports, 30
Pitting, 311, 325, 331, 357, 379-385, 398
Plastic deformation, 155, 163
Platformer heat exchanger, 312
Plug-score groove, 114

Porosity
  castings, 131
  welds, 86, 247-250, 279
Postheat treatment, 305-316
Preheat treatment, 275, 302-305
Pressure cycling, 336
Pressure reducing pipe sections, 30, 295
Pressure vessel failures, 5, 148-150, 299, 339, 351, 353
Proof test failures, 190
Purging, 285, 290

Quench aging, 317
Quenching and tempering heat treatments, 309

Refineries, 53, 71, 72, 284, 312, 355, 405, 413
Rehabilitation, 63
Reinforcement angle, 263
Reinforcement shape, 263
Reinforcing ribs, 29
Reinforcing rings, 25, 29
Restraint, 13, 14
Robertson test, 169
Root-pass cracking, 251, 274

Sand blasting, 365
Scabs, 105
Scale, 131, 325, 330, 356, 403, 416
Schaeffler diagram, 231
Seams, 107
Sensitized stainless steel, 377
Shear fractures, 156
Shielded metal arc welding, 297
Shot blasting, 327-328, 330
Shrinkage cavities, 132
Shrinkage of welds, 390
Sigma phase in stainless steels, 71-72
Sink, 223, 250-253
Slag inclusions
  castings, 131
  welds, 253-256, 260
Slivers, 2, 105
Slow bend notch tests, 168
Slugging, 256
Socket joints, 23, 210
Soil corrosion, 415

Solution heat treatment, 63, 66, 307-308
Sprinkler piping, 335
Stabilization heat treatment, 63
Stacks, 149
Stainless clad steels, 25, 392
Stainless steel liners, 346
Stainless steels
  "H" grades, 308
  Type 304 or 304L, 66, 71, 73, 89, 101, 133, 201, 204, 308, 326, 348, 375, 383, 387, 393, 397, 407
  Type 310, 71
  Type 316, 316L or 317, 32, 69, 141, 236, 349, 377, 407
  Type 321, 69, 236, 308, 339, 391
  Type 347, 67, 76, 119, 142, 146, 215, 230, 235, 287-289, 311, 390, 392, 393, 405, 416
Steam power plant boiler tubing, 91, 92, 95, 107, 200, 281, 308, 310, 311, 327, 330, 331, 352, 380, 382, 398, 403, 405
Steam power plant piping, 17, 25, 29, 53, 67, 69, 94, 104, 107, 142, 283, 342, 343, 373, 403, 413
Steam sampling lines, 15, 390
Steels
  carbon, 51-53, 104, 148, 151, 203, 208, 243, 258, 265, 303, 343, 352, 353, 372, 379, 403, 415
  1 Cr–$\frac{1}{2}$ Mo and 1$\frac{1}{4}$ Cr–$\frac{1}{2}$ Mo, 39, 120, 123, 136, 274, 283, 305, 310, 315, 330, 348, 352, 353
  2$\frac{1}{4}$ Cr–1 Mo, 119, 287, 289, 343
  5 Cr–$\frac{1}{2}$ Mo, 135, 138, 143, 198, 312
  $\frac{1}{2}$ Mo, 30, 53, 85, 94, 140, 142, 145, 221, 304, 335, 344, 413
  nickel, 148, 234
  vanadium, 64, 306
Strain aging, 181
Straining, 151, 155, 180, 189
Stress concentration factor, 26-29, 153
Stress concentrations, 20, 151, 174
Stress-corrosion cracking, 310, 311, 317, 327 387-398
Stress-enhanced corrosion, 312, 406-409
Stress oxidation, 2
Stress relief heat treatments, 95, 183, 311

Index

Stringers, 98
Structural defects, 9
Stubbing, 256
Submerged-arc welding, 233, 273, 286-287, 293, 297
Suck-up, 250
Sugaring, 245
Superheater tubing, 308, 352, 353
Swedged ends, 202

Tank failures, 17, 293
Thermal fatigue, 12, 14, 25, 26, 29, 211, 293, 315, 335, 340-352, 357
Thermal shock, 340
Thermometer wells, 5, 23, 76
Thresher, 6
Titanium, 289-290
Toe cracks, 236
Tong marks, 84
Tongues, 105
Tool marks, 113
Toughness, 4
Transition temperature, 151, 152, 157, 159-161, 176-189, 317
    effects of composition, 176
Transportation, 85-86

Tungsten inclusions, 256-258
Turbine failures, 5, 17, 307, 344
Turbining, 325, 328

Underbead cracking, 221, 236, 302
Undercut, 259-260, 390
Unfused chaplet, 139

Valves, 62, 63, 132, 136, 139, 140, 142, 212, 290, 345, 410, 413
Vanadium steels, 64-66, 197, 306
Vibrations, 149, 337, 339, 347, 357, 398

Wagon tracks, 260-262
Wall thickness variations, 19
Water tanks, 17, 293
Weld crater, 209
Weld defects, 214-271
Weld joint design, 34-47
Weld joint fit-up, 207, 290, 382
Weld reinforcement, 242, 263-266
Weld repairs, 143, 149, 298-299, 312, 331
Weld ripple, 2, 215
Weld spatter, 266, 413
Worm holes, 247, 279